Technologies of International Relations

Carolin Kaltofen · Madeline Carr
Michele Acuto
Editors

Technologies of International Relations

Continuity and Change

Editors
Carolin Kaltofen
Department of Science, Technology,
 Engineering and Public Policy
 (UCL STEaPP)
University College London
London, UK

Michele Acuto
University College London
London, UK

Madeline Carr
Department of Science, Technology,
 Engineering and Public Policy
 (UCL STEaPP)
University College London
London, UK

ISBN 978-3-319-97417-0   ISBN 978-3-319-97418-7   (eBook)
https://doi.org/10.1007/978-3-319-97418-7

Library of Congress Control Number: 2018953414

This Palgrave Pivot imprint is published by the registered company Springer Nature
Switzerland AG
The registered company address is: Gewerbestrasse 11, 6330 Cham, Switzerland

Contents

Notes on Contributors

Michele Acuto is Chair of Global Urban Politics in the Faculty of Architecture, Building and Planning at the University of Melbourne. He is also Senior Fellow of the Chicago Council on Global Affairs and of the Bosch Foundation Global Governance Futures Programme. His work focuses on the intersection of global and urban politics.

Barry Buzan is Emeritus Professor of International Relations at the London School of Economics and honorary professor at the University of Copenhagen and Jilin University. His work focuses on the English School of International Relations, the global transformation to modernity and International Security Studies.

Malcolm Campbell-Verduyn lectures in the Department of Political Science at the University of Toronto. He completed his Ph.D. in International Relations at McMaster University and a SSHRC Postdoctoral Fellowship at the Balsillie School of International Affairs. His research focuses on the roles of emergent technologies, non-state actors and expert knowledge in contemporary global governance.

Madeline Carr is Associate Professor of International Relations and Cyber Security at University College London and the Director of the RISCS Institute for research into the science of cyber security. She is also the Director of the Digital Policy Lab which supports policy making to adapt to the pace of change in society's integration of digital technologies.

Constance Duncombe is Lecturer in International Relations at Monash University, Australia. She received her Ph.D. in International Relations at University of Queensland, and has worked as Postdoctoral Research Fellow in the Faculty of Humanities and Social Sciences, and in the School of Political Science and International Studies at UQ. Her research focuses on the challenges associated with conceptualizing the political power of recognition as it relates to foreign policy.

Myriam Dunn Cavelty is Senior Lecturer and Deputy for Teaching and Research at the Centre for Security Studies (CSS), ETH Zurich, Switzerland. Her research focuses on the politics of risk and uncertainty in security politics and on changing conceptions of (inter)national security due to cyber issues.

Toni Erskine is Professor of International Politics and Director of the Coral Bell School of Asia Pacific Affairs at The Australian National University. She is also Associate Fellow of the Leverhulme Centre for the Future of Intelligence at the University of Cambridge. Her research interests include moral agency and responsibility in relation to both formal organisations in world politics and forms of artificial intelligence.

Yale H. Ferguson is Professorial Fellow, Division of Global Affairs, Rutgers University-Newark and Emeritus Distinguished Professor of Global and International Affairs, Rutgers University. He has published over 60 articles/book chapters and 12 books, including (with R.W. Mansbach) *Globalization: The Return of Borders to a Borderless World?; A World of Polities; Remapping Global Politics; The Elusive Quest Continues; Polities; and The Web of World Politics.* He has been three times a Visiting Fellow at the University of Cambridge (where he is a Life Member at Clare Hall), also Senior Fellow at the Norwegian Nobel Institute, Visiting Scholar at University of Padova (Italy), and Fulbright Professor and later Honorary Professor at the University of Salzburg (Austria). He is the recipient of the International Studies Association's Distinguished Scholar Award (IR and History), and the ISA-NE created an annual book award in his name.

Stefan Fritsch is Associate Professor of International Relations in the Department of Political Science at Bowling Green State University, Ohio, USA. His research examines the impact of technology on global affairs and the comparative political economy of innovation between national economic systems.

Blayne Haggart is Associate Professor in Political Science at Brock University. Blayne received his Ph.D. from Carleton University. He joined Brock in 2012 in the International Relations subfield, and his current research focuses on intellectual property and the rise in government and commercial surveillance on the global economy.

Jonas Hagmann is Senior Researcher, SNSF Ambizione Fellow and Senior Lecturer at the Institute of Science, Technology and Policy (ISTP), ETH Zurich, Switzerland. His research focuses on the political sociology of international risk and security politics.

Carolin Kaltofen is Post doctoral research associate in 'Science Diplomacy' in the Department of Science, Technology, Engineering and Public Policy (STEaPP) at University College London. Her research is located at the intersection between the study of science and technology, digital humanity and politics. She is particularly interested in questions of agency, ethics, corporality and crime in relation to new technologies.

Keith Krause is Professor at the Graduate Institute of International and Development Studies in Geneva, Switzerland, Director of its Centre on Conflict, Development and Peacebuilding (CCDP), and Programme Director of the Small Arms Survey, an internationally recognised research centre NGO he founded in 2001.

Sarah Logan is Postdoctoral Research Fellow at the University of New South Wales Law School. Her research interests include the geopolitics of technology, the concept of the information state, and the impact of technology on intelligence practice. She holds a Ph.D. in International Relations from the Australian National University.

Linda Monsees is Postdoctoral Fellow at the Goethe University Frankfurt. Prior to that, she worked at the Center for Advanced Internet Studies and the Bremen Graduate School of Social Sciences. Her research focuses on networked technology, especially digital encryption.

Can E. Mutlu is Assistant Professor in the Department of Politics at Acadia University, and an expert of international security, technology, mobility, and methods. Can has a Ph.D. from the University of Ottawa.

Joseph Nye is University Distinguished Service Professor, Emeritus and former Dean of the Harvard's Kennedy School of Government. He is a Fellow of the American Academy of Arts and Sciences, the British

Academy, and the American Academy of Diplomacy. He is an expert of power, interdependence and American foreign policy.

Tony Porter is Professor of Political Science, McMaster University, Hamilton, Canada. He is currently conducting research, funded by the Social Sciences and Humanities Research Council of Canada, on 'Numbers in the changing fabric of global governance'. He is also co-organizing with Netina Tan a project on "digital democracy" which explores the positive and negative impacts of digitization on democracy.

Christian Reus-Smit is Professor of International Relations at the University of Queensland and a Fellow of the Academy of the Social Sciences in Australia. He is co-editor of the *Cambridge Studies in International Relations* book series, the journal *International Theory*, and a new multi volume series of *Oxford Handbooks of International Relations*.

Mark B. Salter is Professor at the School of Political Studies, University of Ottawa, and Editor-in-chief of *Security Dialogue*. His research focuses on Critical security studies and International relations theory, questions of technology, securitization, materiality, border and airport security, and actor–network theory.

Saskia Sassen is Robert S. Lynd Professor of Sociology at Columbia University and Member of its Committee on Global Thought, which she chaired till 2015. She is a student of cities, immigration, and states in the world economy, with inequality, gendering and digitization as three key variables running through her work.

Susan Sell is RegNet Professor in the School of Regulation and Global Governance at the Australian National University. She earned her Ph.D. in Political Science at the University of California, Berkeley. Her work focuses on the global governance of public health, food sovereignty, education and climate change.

J. P. Singh is Chair of Culture and Political Economy, and Director of the Centre for Cultural Relations at the University of Edinburgh, editor of the journal *Arts and International Affairs* and also of Stanford's book series on *Emerging Frontiers in the Global Economy*. His work focuses on questions of power, international cultural policy and how technologies can foster inclusivity in the developing world.

Leonie Tanczer is Lecturer in International Security and Emerging Technologies at University College London's Department of Science, Technology, Engineering and Public Policy. She is affiliated with UCL's Academic Centre of Excellence in Cyber Security Research and former Fellow at the Alexander von Humboldt Institute for Internet and Society in Berlin. Her work spans a wide array of Internet-related topics, ranging from online sexism to digital censorship and surveillance.

Ole Wæver is Professor of International Relations at the Department of Political Science, University of Copenhagen, founder of CAST, Centre for Advanced Security Theory, and Director of CRIC, Centre for Resolution of International Conflicts. His current work focuses on updating the foundations in philosophy of language for securitization theory and take research towards a new focus on technology.

CHAPTER 1

Conversations on Technology and IR

Michele Acuto, Madeline Carr and Carolin Kaltofen

Abstract The introduction to *Technologies of International Relations* sets the scene for the volume's 'conversations' between different generations of International Relations theory (IR) scholars. It outlines the growing centrality of technological questions to the investigation of international relations and world politics, the presence of a tradition of engagement with technology throughout the 'classics' of IR, and the need for a renewed discussion on this theme at the core of the discipline. Building on a report from two distinguished scholars panels at the International Studies Association in 2015 and 2016, the chapter advocates for a renewed approach to technology that strides a thoughtful mediation between the heights of past achievements and the current generation of creative and critical thought.

Keywords Technology · Generations · Dialogue · International Relations theory · New Materialism

M. Acuto (✉)
University of Melbourne, Melbourne, VIC, Australia

M. Carr · C. Kaltofen
Department of Science, Technology, Engineering and Public Policy (STEaPP),
University College London, London, UK

1

A Dialogue on Technology
and the Role of a Discipline

In seeking a home for these conversations, several of us took an active duty in the Association to participate in the establishment of an ISA special section on Science, Technology, Art and International Relations (STAIR). Since its 2015 inauguration in New Orleans, STAIR has been hosting a wealth of panel-based discussions at ISA's Annual Convention (ISA) concerning the role and place of technology in International Relations (IR), giving an indication of just how central this debate is to the discipline. That year we set out to have a more explicit engagement with the epistemological and ontological status of technology in IR theory, which in part was also an effort to continue, convene, and document conversations that had become far too critical and far too many to remain without representation in the official programme. Much of what is discussed in this volume is inspired by the numerous memorable encounters in the halls of ISA, and which regularly extended into the afterhours, sparking a series of heated debates that were hosted by pubs and bars off-site. We would like to stress that *The Technologies of IR*, like many other collaborative projects, draws on and is shaped by the intellectual voltage generated at the edges of the programme.[1] Eventually, this endeavour took the format of an annual 'distinguished roundtable' for the STAIR section (New Orleans, February 2015; Atlanta, March 2016).

The Technologies of IR volume is rooted in these informal as well as formal discussions. The roundtables drew in a star cast of scholars like John Mearsheimer, Dan Deudney, Laura Sjoberg, Chris Reus-Smit, Joe Nye, Patrick Jackson, and Barry Buzan, which underscored that the appetite for technology was neither niche nor of the "next generation". Squeezed into two very crowded hotel rooms, the panellists took turns at arguing for or against the value of technology in rethinking their works. Their reflections, objections and confessions are the kind of insight we sought to explore further in this collection, carrying forward 'continuity and change' as our central theme. Apart from framing IR's engagement with technology, the purpose of the theme is to promote a more historically alert engagement with well-established traditions of studying technology, especially Science and Technology Studies (STS) and the Philosophy of Technology, while provoking innovative, bold and reflexive ways of theorizing technology in international relations. In making 'continuity and change' the Leitmotif in the development of this volume, we aim to inspire a greater appreciation of the value and purview of thinking IR through technology.

Put differently, if IR has fallen short of the achievements of other disciplines, this is perhaps more explicitly in reflecting systematically and in a non-determinist fashion on pervasiveness of technologies in the historical evolution of international relations. How have approaches to the relationship between technology and concepts like power, security and global order developed in IR? To what extent can our ideas about the role and implications of industrial age technology help us understand the challenges and opportunities of the information age?

In the pursuit of making a case for technology as central to IR, we convened, challenged and chased IR scholars, students, sources and materials across generations, traditions and research interests, gathering a variety of voices that many would appear largely unrelated. However, the volume much rather reflects an unconventional IR collective that emerged in the conversations between the volume's contributors, roundtable panellists and STAIR-minded co-conspirators. Trying to avoid detaching our sense-making and retelling of these exchanges and insights from the roundtables highlights we seek to capture, this volume is shaped by the explorative and fervent conversations that ensued and those that continued in the aftermath of the 2015 and 2016 roundtables as much it is by the more prolific original events.

In this sense, the interviews on the technologies of IR have a dual purpose; they serve to emphasize the original works that paved the way for us to ask questions about technology as a discipline in the first place, while also providing a flavour of the types of research prevalent to IR as a discipline-yet-to-come. The STAIR sessions continue to inspire explorative projects that are so central to constituting a discipline that is grounded in the deeper considerations and concerns of its younger generations, notably often through the creative engagement with rather than expense of work that has long been distinguished. While generating momentum and increased interest in STAIR matters, initiatives like *The Technologies of IR* make questions of socio-technical analysis, new materialism and posthuman thought a frequent part of the scholarship driving the research and study of IR. Yet, against the canons of innovativeness and the voracious demand for finding 'gaps' in the scholarship, we argue that the growing enthusiasm and interest in studying the relationship between technology and world politics is not developing out of any particular gap.

REDISCOVERING IR's AFFAIR WITH TECHNOLOGY

IR is (re)discovering technology or rather uncovering that technology is no novelty: whether as abstract concept or concrete object of study, technology in its diversity has always been central to IR scholarship, especially IR theory. The pages of some of the most famed and foundational texts of the discipline are crowded with military machines, means of transport, communication and connectivity infrastructures. From *The Twenty Years Crisis*, to *The Anarchical Society* and *Social Theory of International Politics*, nearly all of IR theory has dealt in some measure with the question concerning technology in world politics. In light of the heightened interest in technology currently sweeping across IR, the suffix "(re)" is critical in a number of ways. It serves to acknowledge the historic legacy and expertise of IR scholars in dealing with the 'politics of technology', while flagging the subsequent logic of our approach, the recovery of technology.

The juxtaposition of foundational IR voices in those 2015 and 2016 discussions demonstrates that technology has not only always been present in IR scholarship and fundamental to the discipline's development, but, more broadly, IR's engagement with technology over the last five to ten years raises the question of why technology had not been problematised much earlier. In this sense, this volume is not a case of marking the discovery of new IR territory, neither of closing the gap with findings fresh from the field nor the "next generation's" call to conquer the conceptual wasteland of the notoriously "understudied". If anything, *The Technologies of IR* project is 'genealogical' in spirit, trying to tease out how the influence of technology has featured in IR scholarship, implicitly and explicitly, at a time when studying technology "wasn't a thing".

Ultimately, *The Technologies of IR* 'looks back' at the role of technology in IR's core texts, and 'looks forward' to the part these technologies will play in the years ahead. We chose the "interview turning conversation" style to start the recovery of technology, which to us is an intriguing variation of the hermeneutical method as the interviewer prompts the author to be her own interpreter. This volume contributes to what is arguably an 'effective history' of technology and IR, which is the continuous effort across generations of IR scholars.

An Intergenerational Dialogue

The 'looking forward' of this volume is also embodied in its purposeful pairing of generations, connecting influential thinkers of IR with the emerging genus of early and mid-career scholars, a generation marked by post-traditional IR and deeply embedded connections with the likes of STS, anthropology, sociology and political geography. Mirroring the Q&As of the 2015 and 2016 roundtables as much as the typical convivial atmosphere of afterhours ISA chatter, *Technologies of IR* is not a book of chapters, which acknowledges foundational figures for contemporary IR by initiating a direct dialogue between and across generations. The interview format provides the space for individuals to reflect on a lifetime of work with, through and on technology as well as to explore new ideas about technology in conversation with early-er career scholars who have taken a 'STAIR' approach to IR and are fast emerging as new exciting voices in IR thinking. The voices that gather in this short volume are, of course, not a representative crowd of IR, or indeed pretending to stand for a generalizable research programme or approach. Rather, as the cordial tone of many interviews flags, this is a series of exploratory conversations that aims to engender more debates and intergenerational dialogue. They do so, we believe, with the respectful and warm spirit flagged above without pointing at obvious previous flaws in IR's consideration of technology in world politics, suggesting that we forge future (cross-)disciplinary advancements *not* in a rupture from the 'past', but in a dialogue.

The style of writing is in itself not new but owes much to the trail blazed by colleagues like Peer Schouten with the project *Theory Talks* and to the supportive and respectful collaboration between several foundational figures of IR thinking and the newer ranks of IR theorists. The central themes shaping the direction of this project emerged in and across their conversation and we will begin this book of discussions by stepping back to the 2015 and 2016 debates.

Debating Tech at the Heart of IR

The Technology of IR roundtables and conversations have dealt with literally decades of IR scholarship and, hence, are too rich to be simply summarized in a few lines of an introduction. We will offer a taster of the conversations instead, hoping that it will be a useful entry point to and overview of the diverse debates that the interviews of this volume touch upon. So where to start?

The Technologies of IR are about introspection, reflexivity and finding new connections. For instance, Chris Reus-Smit pointed out that there is, in fact, a deep story of technology in his landmark *Moral Purpose of State*, which details how the transition from the absolutist state and the shift in the constitutional architecture of the (international) system of states to a modern form is embedded in a technologically driven revolution of intersecting ideas. In a similar vein, Joe Nye highlighted that his early work on transnationalism was ripe with critical shifts in technology and globalization, and John Mearsheimer recognized that his engagement with security and military affairs explicitly flagged how the technology at the core of IR changes constantly and how capturing its relationship with key questions of (international) security is an extremely complicated enterprise. While these conclusions and concerns are not new to those IR scholars who have been taking a specific interest in the role of technology for some time, the point here is to use such shared concerns to establish the study of technology as foundational to the discipline of IR rather than a field of consideration by some scholars of IR.

In this sense, the dialogue-based approach to *The Technologies of IR* has also been about recognizing the limits we put on the study of technology by assigning it to a particular 'field' or 'area' of scholarship. Reus-Smit emphasized this at the 2015 roundtable by noting how technology challenges "our mode of theorizing and the horizon of what we deal with"—a "test for existing sets of paradigms because these might simply not be particularly useful" and so there is "no paradigm talk here, no isms". Nye admitted how his study of power and interdependence, which was widely recognized for its theorizations of complex interdependence, had assumed the role of major technological changes but perhaps failed to discuss them explicitly. He flagged that having to "learn fast" about critically new or pervasive technologies is an increasing challenge for many IR scholars and not only because they were hardly familiar them. Equally, even for scholars that typically look at 'big' trends and 'large' matters, such as John Mearsheimer and Dan Deudney with nuclear security, IR cannot discount that technology is a pervasive affair, and one that can be met in the minute and situated. This is well known by Laura Sjoberg who, in the 2016 debate, took us through her work on feminist security studies in order to flag the importance of understanding the "logics of technology" as embedded, embodied and situated relationships. Hence, and as seconded by Deudney, the "micro-scale is critical: micro geopolitics created by micro technologies" are "equally as central as IR's big questions". We draw from this an 'inter-generational' agreement to move beyond traditional IR approaches. The growing engagement of IR scholars with other disciplines, such as STS, Media and Communication Studies and Criminology, underlines such an approach as well as emphasizes that the relationship between IR and technology cuts across registers, requiring a study that takes technology as seriously as IR and that is open to following technology 'all the way'.

The analytical challenges to taking such an approach emerged early in the discussion and pointed to the shortcomings of 'technology' as a context of study itself. For example, Reus-Smit in 2015 suggests that if we see technology as the application of scientific knowledge for practical purposes, then the very materiality of technology is also embedded into two very social things: ideas and practices—leaving space for constructivist modes of analysis to tell us about the technologies as much as the sciences of IR. This sense of technology as "congealed ideas, not just an apparatus", seeing technologies as inextricably tied to practices—something that "by definition is material and ideational and thus embodies a profound set of theoretical challenges to IR". Adding to the challenges, Patrick Jackson provocatively flagged in his contribution to the 2016 roundtable that our work does not deal explicitly with technology but rather *science*, and that we should treat this as distinct from the 'tech' frenzy in IR. Technology, he reminds us, is *not* science: science is an internally diversified language game, and a standard for claims based upon a series of methodologies (or a "system of statements") which is removed from technology insofar as science is "stepping back from a fact or the object of studies". The S in STAIR was in this sense no casual addition and within the remit of this project critically complements the study of *The Technologies of IR*. However, science and technology are routinely conflated or collapsed into "science *and* technology". While always unhelpful this is rarely acknowledged and addressed, and we too were careful not to reproduce this habit of thought in our discussions. We want to emphasize the difference between science and technology as a critical element in IR's technology puzzle, which several dialogues in this volume surface quite clearly. Moving forward, *The Technologies of IR* research agenda, therefore, also needs to develop a deeper consideration of the relationship between technology and science itself, which we argue offers a key steppingstone for an IR intervention into the role of technology in world affairs.

So, what does IR bring to the table when it comes to studying its relation to technology? The Q&A debate from the 2016 roundtable is telling here. To begin answering this very question Jackson cautioned that IR as a whole might not be able to "bring" anything as it is in fact "not a thing"—it is in the actions of various embedded scholars that bring different kinds of things to the table of understanding technology that the bigger value might lie here, and we should perhaps recognize how fundamental questions such as those of technology show us how we "worry too much about what the essence of IR is". Sympathetic to Jackson's appeal Sjoberg, addressed this lack of unitary contributions as a good thing: "there isn't one IR way of looking at something," but it is also true that the "tools and concepts of paradigmatic IR can well normalise technology," thus, making it graspable and analysable from specific viewpoints. So, does Deudney, noting that the study of technology in IR is confronted by the "challenge of fragmentation" and what might be a "misplaced expectation of the emergence of shared understandings". However, as Sjoberg pointed out then the emphasis on 'innovation' as a key feature of technology also presents the theoretical problem "of something new pushing out of the stage the 'old'"—a bias much in need of redressing.

We stressed the difficulty of convening various scholars to initiate 'new' research without juxtaposing it to an 'old' earlier, which is a challenge to be taken seriously, in terms of both the tireless developments of technology but also our intergenerational dialogical approach. The concern that traditions of valuable insight are too readily traded for the glitzy new is echoed far across the spectrum of IR by voices such as Mearsheimer, who, following suit to Sjoberg's comment, remind the audience that IR materialist theories (e.g. Realism, Structural Marxism, economic interdependence theory) do "put a premium on technology" and should be acknowledged for that, not ditched in favour of the theoretically new. What is more, some scholars among the materialist traditions in IR have long been acknowledging the centrality of technology in the 'human condition', development, security and culture. As Deudney argues in the first roundtable, "we don't need fundamentally new approaches: old questions are still relevant" because traditional IR thinking embodies a "set of arguments on violence and interdependence in anarchy which still matter a lot" especially as technological advances open up the possibilities to de-stabilize world politics.

The general apprehension that studying technology in IR could result in 'straw manning' and dismissing traditional IR undoubtedly warrants closer attention. Yet, this is also because the concern itself encourages the stereotyping of IR scholarship on technology as opposing well-established traditions. Given that much of IR's 'new' treatment of technology builds on the intellectual legacy of the likes of Martin Heidegger, the Frankfurt School, Jean Baudrillard and Donna Harraway, just to name a few, raises questions about the role that the study of technology has taken in IR so far. More specifically, these are about the ways in which traditional IR has engaged with the work of later generations as well as about the appropriateness of framing this challenge of addressing technology in IR as a tension between old and new, as change without continuity. *The Technologies of IR*, like many other areas foundational to IR, has to take the "pushing out of the stage" seriously not just in case of the old, but even more so in that of unconventional, non-traditional or non-mainstream IR scholarship.

In this sense, Nye foregrounding that IR theorists bring meta-theory to the table is highly applicable and critically fundamental to the reflexivity of any theory in IR for it builds our capacity to grapple with vastly different meta questions about power, stability, and sovereignty. For Nye and Buzan it "takes a long time for a technology to be pervasive" and, thus, IR's capacity to look at the 'long duree' of technological transformations remains a traditional but highly valuable contribution, which is, for instance, embodied in the assessment of both nuclear and digital threats. This is not an invitation to sit comfortably on our IR skills, however, as there is a vast "variety of technological wildcards that could change the game of IR," to use the words of Buzan. Therefore, and as suggested by Nye, "don't fear technology, plunge in and try to understand, because IR theory can provide as equally valuable lessons as many other disciplines". What is needed, rebutted Reus-Smit, is to understand that the challenge for IR should be to step beyond the curious and into the empirical advancement—"not just 'add technology and stir', but a program of research" capable of shaping IR thinking for the long run.

A FIELD REPORT ON THE TECHNOLOGIES OF IR

The Technologies of IR volume is looking back and looking forward—a collective effort to join the history of IR thought and current international theorizing of technology, which is captured in the pages that follow. As editors we see riches in these conversations not only because they convene the reasonings of some of IR's major thinkers, but much rather because they present us with a horizon of analytical possibilities. Collectively their work reflects the 'depth' of the role of technology in IR, which is a challenging and exciting time to experiment with the study of technology in international relations as well as a serious task ahead. Hence, this is not an attempt to "add technology and stir" in order to think IR anew. Rather, this diverse series of conversations proposes an IR thinking that understands the role of technology through the thoughtful mediation between the heights of past achievements and the current generation of creative and critical thought. The interviews in this volume mediate by tracing technology's ubiquity and influence in the foundational texts of IR. Rather than "pushing out of the stage the 'old'" *Technologies of IR* builds an edifice of STAIR-like theorizing that is based on dialogue. This also means, however, that we are neither offering nor aiming at a definitive statement on what the theory of, approach to, role or place of technology *is* in international relations, whether in practice ('ir') or theory ('IR'). In this sense, we do not come to a traditional 'conclusion' of a new '-ism' or paradigm. Quite the contrary, the goal of the next pages is to showcase that IR thinking through and about technology will always be in tension, that is either debated, advanced, disputed, abandoned, recovered, reinvented or theorized anew.

The extraordinary theoretical complexity of technology is reflected in the thinking of the classic 'textbook authors' that have come to shape the discipline, as much as in the observations and queries of their students and interlocutors, a group of fast emerging scholars that exhibit a nuanced flair for technology and its role in IR. As our introductory discussion optimistically evidenced, the diversity of ideas is underpinned by common cross-cutting themes or concerns about the role and study of technology in IR, such as taking the implications of technology for ir and in IR seriously, and the value of intergenerational and cross-disciplinary modes of thinking about world politics in understanding *The Technologies of IR*. Hence, we see this volume as being a promising field report, rich in evidence and detailed insights, curious quotes and inspirational thoughts. We hope that observing the conversations around *The Technologies of IR* offers thoughtful and creative answers, while also engendering new discussions and questions. Ultimately, we want to stimulate more of that intergenerational dialogue that makes for a healthy enduring discipline.

NOTE

1. On a much more fundamental level, this is an argument for the critical role of the well-guarded academic privilege of being "Out of Office" to engage "face-to-face" in a traditionally random fashion, after just having bumped into or while rushing together in the crowded hectic of the "in-between".

CHAPTER 2

Theory Is Technology; Technology Is Theory

Linda Monsees in conversation with Ole Wæver

Abstract New technology is undoubtedly changing world politics. But does this necessarily require new theories? In this interview, we explore the challenges facing a (political) theory of technology and how to understand the novelty of technologies such as Big Data. Ole Wæver recounts his early interest in technology and how theorizing technology demands that we look at different kinds of acts. Some of the main challenges include unintended effects and the assessment of decisions made within complex systems. We go back to Langdon Winner's early work on the political character of technology, and discuss why his ideas might be more valuable than concepts often subsumed under the heading of 'New Materialism'.

Keywords Theory of technology · New technology · Politics of technology · Big Data · Critique · Power

L. Monsees (✉)
Cluster of Excellence Normative Orders, Goethe University Frankfurt, Frankfurt, Germany

O. Wæver
Department of Political Science, University of Copenhagen, Copenhagen, Denmark

C. Kaltofen et al. (eds.), *Technologies of International Relations*,
https://doi.org/10.1007/978-3-319-97418-7_2

13

I met Ole Wæver at the annual ISA convention in 2016. While I wasn't surprised to see him at a panel on realism, I got curious when he kept showing up at virtually every session on new materialism, technology and ANT. Eager to find out why he was interested in these discussions, I figured that my best chance of finding out what he was up to was by asking him for an interview. He was generous, agreeing to meet me in spring 2016. We met in his office in Copenhagen discussing technology, theory and the current challenges for IR. This is not surprising given that Ole Wæver is most known for his theory building in security studies. His ideas on the role and function of theory show up during the interview (Wæver, 2009); for both of us the core question is how to theorize technology in IR and what we can (not) learn from STS. We also discussed his general stance towards the theoretical challenges posed by technological developments such as AI. In outlining his ideas on 'theory as technology' the interview revealed the link between his previous work on the theory of security politics and his current project on technology as theory, which due to its ambition leaves me yet again very curious to find out about his future work.

Linda: On reflection, how do you feel technology has been present in your work? What kind of influence has it had on your thinking about IR?

Ole: Well, that is actually the weirdest thing. I was interested in technology, before I started doing any of the things anybody ever heard of. I spent most of my time as a student writing with two fellow-students the equivalent of a BA-thesis, this big thing on philosophy of technology and political technology; 250 pages inspired mostly by the writings of Langdon Winner (Andersen, Wahl, & Wæver, 1983). That is kind of my secret past and that is what I originally worked on.

This is ancient [on his desk lies the old manuscript written with a type-writer], as you can see, before people like us had computers. I wrote about political technology and the implications of computers and information technology at a time when it wasn't pervasive yet (1984). Then I wrote my MA thesis a year later doing something – a conceptual analysis of 'détente' – that pointed much more in the direction of what I have been doing since.

So, in that sense I always had that interest in technology in the back of my mind. But clearly it hasn't played a big role in anything I have been doing. I've been writing about international affairs in a broad sense where of course technology is a part of big power politics. But

I never foregrounded technology. Securitization theory as well does not particularly talk about technology. It is not something I've made explicit at any point. But I want to do it now. I am coming back to the issue of technology at the moment (Wæver 2014, 2017).

Linda: Was your work at that time already within political science?

Ole: That is a good question (laughs). We thought it was, but our teachers thought it wasn't *[LM: Mine neither.]*. We got a very bad grade because they basically said this is not political science. I had a very good teacher in sociology of science who inspired me to do the sociology of science stuff I have been doing in IR (Wæver, 1998). He was our supervisor and our study drew on sociology of science, studying technology through that and thinking about this as politics. We thought that should be part of political science, but the department did not think so at that time. Maybe they (now: 'we') would think differently today.

Linda: Now, you are coming back to this topic of technology. Do you think that there is a fundamental difference between technology of the industrial society and technology of the information age? To what extent do you think can we observe really new phenomena?

Ole: I do think there are some really new phenomena. Some of them are currently researched, but some of them are curiously out of sync with current research. I think we have all along underestimated technology and haven't theorized technology *as* technology very well. I hope we get back to that. I think a lot of the hype that has been about science and technology studies and now new materialism misses the most important side of technology which is in my view the *politics of technology and technology as political acts.* Now, everyone has to upgrade their understanding, not because of a shift from industrial to digital technology, but rather because of the increasingly autonomous actions of technology. The whole artificial intelligence tendency for systems to increasingly act on themselves is kind of a revolutionary change. It is not the digital as such, but Big Data, artificial intelligence, all these things where we increasingly have no longer the ability to make technology transparent. We can't say what does the technology do.

We can say what we put into the machine, but it is no longer the case that one input will produce a specific output. The technology has taken so many steps on its own after whatever we put in, that what it does is beyond anything that can be backtracked to human decisions. That is new. And that is now.

Linda: Talking about autonomous technology or artificial intelligence, does that mean that you would agree with the new materialist notion of distributed or diffuse agency?

Ole: Yes, in a very abstract sense. I am not sure if it is the most helpful way of talking about it. I am not so sure if it is helpful to place the main observation in such an abstract generalization. In some sense it flattens the story. The notion of diffuse agency puts agency in many places and then we might overlook the particular nature of those technologies that have the strongest independent capabilities. That happens if we talk in too general terms.

Linda: *If we want to get at phenomena caused by these new technologies that you just described, do we need a new kind of theory? Your work has also focused a lot on the concept of theory, so what kind of theory do you think is needed here? It seems that current work in IR is focused a lot on new phenomena, but there seems a lack of new forms of theory.*

Ole: I think that is right. We really need to think about how to think about theory. It is important to make quite precise and targeted theories that zoom in on particular dynamics of phenomena that we are interested in and expose these specific mechanisms that are relevant. Just as securitization theory does it. Some people read it in kind of a third-hand textbook version that securitization is just saying something about a rhetoric of fear and scaring people into buying whatever the politicians want to do. But that is not the theory. The theory is much more precise by having some specific concepts and, thereby, getting at some particular dynamics and mechanisms. And that is the kind of theory we need for technology which grasps some of its peculiarities. I am not convinced yet that the new materialism terminology is the most helpful in that. To me it seems like it might have an opening effect because it allows us to speak about things that certain other ontologies would exclude, notably those that operate rigid distinctions between social and material. It is helpful to break some barriers because it becomes possible to say new and important things. But it is not preferable to move from one rigid problematic distinction that isolates agency for humans to a general lack of distinctions; better would be one that specifies different kinds of agency available to types of things and beings. I think as a way of addressing things, we have to be more old fashioned by getting the technology *as* technology back in. 'Good old sense' means here that what technology is has to do with tools and instruments and things you can do things with. The productive and enabling nature of it gets a little bit lost in the new materialism vocabulary. We need to get more into what is done, who is doing what and what is doing what. Too often, nowadays, technology is assimilated into 'matter' (that matters) and 'things' (that act); but even if matter and things should be upgraded, technology can still do things that other things can't.

Linda: *I have the feeling that you really want to do a new project independent of speech-act theory. So I don't want to constrain you to that issue. However, I get the impression that there is some consistency in the idea of politics as 'doing stuff', emphasizing its productive character. Is it the concept of politics that holds both themes together?*

Ole: Well, I would say it is about different kinds of 'acts', which makes it possible to link back to speech act theory. It is the question of who and what can do what kind of acts and, especially, the kind of second order acts, which means acting in the ways that makes other acts possible. That is the most important thing in this context of theory and technology. I was thinking when I was preparing for meeting you, that in some sense my slogan for this would be something like: 'theory is technology and technology is theory'. Think about our job as supervisors at a university and we are talking to a student and we say: 'You have to use theory for your project.' And then the student says: 'What does that mean?' The point we are making then is that using theory means to explicate what difference did it make to your project? What could you do that you couldn't otherwise do. And what is it that becomes impossible by using this theory? This is very much like technology. For me to think about theory is thinking about it as a technology.

I wrote two articles on securitization theory where I try to think about 'what is theory?' (Wæver, 2011, 2015). There the dilemma is how do you think politically not about using theory, but about producing theory? The political assessment of theory becomes too often something like: 'This is a progressive theory because it says some nice things and the picture of the world is so and so. And this is a bad theory because he talked about something bad or there is some insulting stuff in it'. But whether a theory is good or not, progressive or not is ultimately about how it limits or opens up what you can do with it. The good-doing is not what you are doing with it yourself, but what others are doing with your theory. For me as someone who makes theory, the ethical self-reflection has to be about how you build into the theory certain things you can do or can't do, certain things you are almost imposed to do.

If you use securitization theory you have to politicize whatever it is you are dealing with because it makes more explicit the politics involved. In that sense, theory, in essence, is an act of a very special kind because the act acts beyond your control. And this is the similar question you have to ask with technology. It is not only what you do, but what it then can do on its own. And that is different from the bigger emphasis on the experimental side of it. That is where the whole STS theories have been productive but also problematic because it has tended to show the social impact on science and technology. In the end, it reduced technology to an object like any other. Ironically, the result was that everything is socially constructed, there is nothing privileged, there is no secret truth in the essence of natural science. Technology does not correspond to reality but is socially shaped. Science and technology are constructed like everything else. And I think that is why STS gradually lost the ability to grasp the 'technologiness' of technology. There was in recent years a certain coming back to seeing what is peculiar of science compared to other practices. But in the case of technology I don't think STSers have performed the same rediscovery.

Understanding this 'technologiness' is a delicate move that has to face up to a likely charge of technological determinism. Earlier pre-STS technological determinism said that it is because of the laws of nature that technology will take a certain form and, therefore, can do certain things. When STS then wanted to show that it is not nature speaking through natural science or technology, but it is socially shaped then they risk creating the opposite picture. Ironically it ends up being social through and through. And thereby everything is the same, everything in the world is just crisscrossing objects shaping each other. But you *do not* get the peculiar nature of technology as something that can act in a special way. Making technology is similar to making theory. Theory has to be viewed different from first order statements, because theory is a tool for making research and observations; similarly, technology should not be conflated with 'ordinary' objects. Technology for me is objects squared or objects in a second order. It is great to have all objects 'elevated' relative to humans, but too bad to have technology 'lowered' to the level of all other objects as part of that liberation.

Linda: I agree with what you said in some cases of Actor Network Theory, but the STS movement as such is much broader … And I am not sure if your critique applies to all of them.

Ole: No, I think it does. Other non-ANT streams within STS do not say it in these terms, but they are doing it in effect as well. In essence, they do it by fighting against any privileged claim on the behalf of science and technology. You end up creating a universal theory where everything becomes alike. The discourse theory does the same thing. Discourse theory makes it often sound like as if everything is the same. It does not matter if you do a discourse analysis of science or a parliamentary debate or teaching or whatever. It is all discourse; it is all the same thing that is happening. But you lose the ability to see the peculiar nature of religion or of science. And I think the same happened not only to ANT, but to the whole STS 'movement' in the broad sense.

Linda: *Still, I think that the STS people such as Langdon Winner* (Winner, 1980) *whom you mentioned were much more open to politics. I think that IR can actually learn more from these earlier contributions within STS, since they do not make the mistakes you just described.*

Ole: Exactly, but Winner made a critique of STS (1993). He felt that he was among the old philosophers of technology before science and technology studies came. And there is the sentiment that they held on to the critical reflection. They were not the big movement; they were a generation before. Within the STS movement, the classical philosophy of technology was mostly continued by post-phenomenologists. True, they are 'housed' within STS at large, but often assigned particular limited roles, whereas STS has silenced the more principled attempt from say Mumford (1934), Ellul (1964) and Winner (1977) to analyse 'autonomous technology' (1977), which becomes increasingly problematic as we are facing – increasingly autonomous technologies! The deficiency is analytical but politically disastrous. If we exaggerate the STS dogma that all technologies retain 'interpretative flexibility', it becomes impossible to assign political responsibility to technologies that deserve it. Drones, AI, geoengineering ... in IR these technologies act in peculiarly technological ways, and we need the analytical and political vocabulary to address this. Pre-STS philosophy of science (for all their flaws) dared talk politics to technologies.

Linda: *So, you are suggesting that we take two turns back.*

Ole: Yes. In some sense, yes. Because of what is in store for us, technologically, politically and analytically.

Linda: *We talked about theories as technology, but you also said that technology is theory. I am still wondering what that actually means.*

Ole: Uh, now I really get in trouble. This is even more reactionary than what I already said, because it is very controversial whether you see science and technology as two sides of the same coin or not. This probably sounds quite old fashioned. Most modern technologies are science-based in some ways. In any case, they contain conceptual constellations, I don't dare say knowledge or something else even more loaded, but they contain a conceptual apparatus build in material in order to be able to act upon other material. So there is still a link between science and technology to some extent, meaning that some kind of constellation of epistemic categories are related in a way which is then built into a technology. Sometimes the technology might take shape before the science – this is not about process tracing - but the technology nevertheless exploits sciencable relationships. Especially strong technologies and their ability to do things are anchored in a conceptual element inside themselves. The computer obviously builds on all kind of theories and categorically structured interventions into that material. The same goes for all other strong technologies, they are based on science. In that sense technology is theory.

Linda: But isn't it interesting that Big Data incorporates this idea that we don't need theory anymore, it is just pure knowledge. We don't need concepts we can say how things really are, that seems a prevailing sentiment within social science at the moment.

Ole: The practitioners don't need us to understand them. Still, there wouldn't be such a thing as a machine processing Big Data if it did not contain theory. It is just that we don't understand its theory. That is why I find this period we are now in really interesting. Because until recently there would always be some humans who could say: 'I have designed this car in that way', and the engineer can say 'this is related to that. I made this experiment, then I adjusted that and, therefore, it looks now like this.' That might have been partially self-delusional already at that time, but that is how we tended to think about it. We now have increasingly technological systems where no human is a full expert on what the system is doing. That already actually evolved during the Cold War. One of the reasons why the systems warning against incoming nuclear missiles were so dangerous was that they were layer upon layer of computers where no one any longer understood the original systems not to speak of the emergent totality. And no one had been able to afford or to risk taking the whole thing down and reconstructing it. So you had been building on top and on top of systems so that we ended up using a system we did not understand ourselves. But we knew how to make it work and we know that this is exactly what is now happening on a much greater scale with Big Data where the system can produce something that works. We can't really trace every step of it because it has self adjusting algorithms that can do stuff. It is not built based on theory that we outside the system understand, a theory which we need in order to speak to the machine. But, of course, the machine has a theory.

Linda: But if you put it that way, politics vanishes!

Ole: Not if you want to engage politically with the machine. I am just saying politics get one step removed or it gets doubled. You have to take political decisions when we are making, designing or allowing these systems. But also the systems themselves have politics in the things they are doing.

Linda: I am not really sure if you can distinguish between the politics of design and the politics of technology, as if the technology itself could really act. I am still skeptical about that.

Ole: But it is! I mean when you go to the airport and to an airplane and the system is saying you're are a risky person you are not allowed in here. Who made that decision?

Linda: Well… I would say someone wrote the code, and the effect is that we think no one is behind the technology. That we think it was a neutral decision.

Ole: But I don't think it is either/or. It is not neutral, nor can it be traced back to the intent. That is the whole argument that the Big Data people are making. Before big data it was: 'I made these criteria. If you fall under category x then you're are refused. But if you are like z, you are allowed.' But now the system is self-correcting and it makes all these assessments based on its success rate and depending on how well it scores, it will start to build effective categories in its operation. And after a certain point you can't any longer pull out of that. You can't really say what are the criteria? No one knows, it just got so good at doing what it is doing that it is running the show. Those decisions that it is making there, we have, I think, to criticize politically in the same way as if it was a human doing it. But it is not, nor is it determined by the input. It has evolved in its own history.

Linda: But I would still insist that this idea that technology development occurs without human intention has some politically negative implications.

Ole: I totally agree. I mean these screening systems and the whole profiling thing is still a brilliant example. Because sometimes it can get even more discriminatory than anything we could stand up for. Because if it was some kind of transparent system you would be able to say to those who control it: 'Do you really think that skin color is a legitimate criterion?' No it is not! Can you put religion in it? No I can't! But when the system is allowed to work the Big-Data-way then it might actually produce the same effects as if you put skin color in. It is just able to get to that result by other routes. Data! Then you can say, of course, that there are lots of human decisions involved steering the system so it gets to producing those results. But it is not one-to-one intentional, as it would be if someone had figured it out. It is a multitude of small decisions that help the system evolve into a form that fits certain purposes and agendas. Still somehow we have to double our critical engagement. We both critically engage with the people that make the decisions that allow a system like this to evolve and gain power. But, you also have to engage with the technological system as it emerges, what it has become. Since there is no longer any method that can anchor the system in its 'human' origin, because it has a historicity and a path-dependent development. It has its own emergent qualities, meaning that it will not be enough to engage critically with it only through its input. And there you have to hold the system itself politically responsible as well.

Linda: So, now in the end we have a theory of technology which proposes not a flat-ontology ...

Ole: Exactly, because all things have degrees of technologiness. But I still think that it is not enough to treat technology as any other kind of object. Technology is object[2]. It is an object that can do things on other objects. We need to learn to engage politically with technologies in ways that differ from the ways we confront both humans and plain objects.

Linda: Okay, so no flat-ontology, you need some space of politics in that theory and you need to rethink human agency in relation to technology ...

Ole: Yes.

Linda: ...So, that is a big task ahead for you....

Ole: Yeah...yes. For me, for us, for you. But, this is only one aspect of what I am trying to do. We have to think about it as a bottom-up approach, but not only. What we have talked about today is so to say the methods side of it; how should we study and theorize technology? But there is also the top-down or the outside-in question of the whole picture. We have to have an understanding of how the world system hangs together. I mean there are certain technologies that are transformative, that are key nodal points in the whole global development. And we have to study that not only by developing ways of looking at it, but also by studying the system as it evolves. So it is *worse...* we have to do even more!

Works Referenced

Andersen, K., Wahl, T., & Wæver, O. (1983). *Prometheus og Panodora: Informationsteknologi som politisk teknologi* (BA thesis, Department of Political Science, University of Copenhagen).

Ellul, J. (1964). *The Technological Society.* New York: Knopf (French original 1954).

Mumford, L. (1934). *Technics and Civilization.* New York: Harcourt, Brace & Company, Inc.

Wæver, O. (1998). The sociology of a not so international discipline: American and European developments in international relations. *International Organization, 52*(4), 687–727.

Wæver, O. (2009). Waltz's theory of theory. *International Relations, 23*(2), 201–222.

Wæver, O. (2011). Politics, security, theory. *Security Dialogue, 42*(4–5), 465–480.

Wæver, O. (2014, January 2). Open science to fight big threats like the nuclear bomb—The centennial of Niels Bohr's atomic model. *The Conversation.* https://theconversation.com/open-science-to-fight-big-threats-like-the-nuclear-bomb-21591.

Wæver, O. (2015). The theory act: Responsibility and exactitude as seen from securitization. *International Relations, 29*(1), 121–127.

Wæver, O. (2017). *The international politics of dramatically dangerous technologies.* Presentation at the annual meeting of 4S, Boston.

Winner, L. (1977). *Autonomous technology: Technics-out-of-control as a theme in political thought,* The MIT Press.

Winner, L. (1980). Do artifacts have politics? *Dædalus, 109*(1), 121–136.

Winner, L. (1993). Upon opening the black box and finding it empty: Social constructivism and the philosophy of technology. *Science, Technology, and Human Values, 18*(3), 362–378.

CHAPTER 3

A Double-Edged Sword?

Blayne Haggart in conversation with Susan Sell

Abstract Susan Sell's academic career is a testament to the benefits that come from transcending disciplinary boundaries. From her graduate school work, on the effects of technology transfer on development, to her continued contributions on the international political economy of intellectual property rights, her desire to understand economic development has incorporated IPE, legal studies, regulatory theory, history and economics, to name only five areas. Despite early criticisms that this was "not really political science," she has been vindicated in her belief that this work "was important and that it wasn't going to get any less important." Intellectual property rights are widely recognized as essential contributors (and barriers) to economic prosperity and national security (Halbert in Inf. Soc. 32: 256–268, 2016). In this interview, Sell discusses intellectual property's relationship to technology and development, the challenges of conducting interdisciplinary research, and whether we're witnessing the emergence of a counter-hegemonic movement against ever-stronger intellectual property rights.

B. Haggart (✉)
Brock University, St. Catharines, Canada

S. Sell
Australian National University, Canberra, Australia

© The Author(s) 2019
C. Kaltofen et al. (eds.), *Technologies of International Relations*,
https://doi.org/10.1007/978-3-319-97418-7_3

25

Keywords Digital technology · Cyber security · Foreign policy · Internet

Susan Sell's academic career is a testament to the benefits that come from transcending disciplinary boundaries. From her graduate school work, focusing on the effects of technology transfer on development, to her continued contributions on the international political economy of intellectual property rights, her desire to understand economic development necessarily incorporated IPE, legal studies, regulatory theory, history and economics, to name only five areas. While she says that "it was frustrating that some people would review my work and say, 'Well, it's not really political science', because it was very inter-disciplinary," at the same time, "I was very convinced that it was important and that it wasn't going to get any less important, and it would probably only get more urgent as globalization proceeded apace."

She was right. The United States has identified intellectual property protection as an issue of national security (Halbert, 2016). Intangible assets, which include intellectual property, account for anywhere from 50 to 84% of the market value of the Standard and Poor's 500 index (Monga, 2016; Ocean Tomo, 2015). Furthermore, intellectual property rights-based monopolies have emerged as a barrier to technological innovation and development. Thankfully, for political scientists (and others) wishing to understand the importance of intellectual property for exercising power in the global political economy, Sell's work provides us with a solid foundation upon which to build. Over the course of an hour-long Skype call in April 2016, we discussed intellectual property's relationship to technology and development, the challenges of conducting interdisciplinary research, and whether we're witnessing the emergence of a counter-hegemonic movement against ever-stronger intellectual property rights.

Blayne: What do you think technology is?
Susan: That's a hard question for me to answer. It's almost funny, because it's so simple, but when I think about technology, I tend to think about capacity. I tend to think about inventions or innovations that allow people to do things. So, I think of it as capabilities, I suppose. It's a double-edged sword—obviously, there's nothing deterministic about

technology, but I have always been interested in the potential for technology to improve people's lives and to promote development: economic development, human development.

Blayne: How do you feel technology has been present in your work?

Susan: Well, I initially got interested in it in graduate school as a topic because I was interested in the question of technology transfer and economic development, and I was doing research on different countries' development strategies and trying to acquire technology. I did some historical work on it, which I found quite fascinating, about how different countries learned by doing and would innovate by importing technology and changing it up.

And when I was looking at the question of technology transfer and development I, of course, was looking at investment rules, I was looking also at intellectual property rules, and came to find out that intellectual property protection can be a barrier to access to technology and that's how I got so interested in intellectual property.

Blayne: When you were starting out, was technology transfer in the IR or IPE mainstream as a research topic?

Susan: Oh, absolutely not. My mentor was Ernst Haas and he was very interested in the role of science and international organizations and he was very supportive of the topic, but it was absolutely out of the mainstream. Very few people—hardly any in fact—were writing about economic development in IR. Very few people that were working in political economy were looking outside of the OECD countries, so I felt very much out of the mainstream.

There were a few scholars at the time working on some of these issues, especially development—that would be Craig Murphy and Thomas Biersteker who were supportive, and Ernst Hass, of course, was always very supportive—but, no, my work didn't speak to the mainstream and the mainstream didn't really speak to my work.

[Over time,] I would argue that intellectual property—which is something that I've focused on throughout—to the extent that intellectual property rules have gotten more intrusive and expensive and really affect our daily lives, people have been paying more attention to it. And I think that both digital protests and live protests against the copyright laws for the [US] *Stop Online Piracy Act* and the *Protect Intellectual Property Act* in 2012 highlighted how much people have come to rely on access to digital information and how engaged they are in paying attention to issues around privacy, in addition to issues around access and, you know, the tragic suicide of Aaron Swartz who was a real activist for making educational materials available is one good example of that.

Blayne: What kind of influence has technology had on your thinking on international relations at large? I mean, Private Power, Public Law *is not just about IPE, it's about how policies are made internationally. It's about the big issues of international relations.*

Susan: I think it [*Private Power, Public Law*] highlights the role of private actors in making policy and the lack of transparency about how the policies are made and how they are made to privilege the few at the expense of the many. I think that's one of the takeaways from that book. But [writing the book] also made me realise how valuable intellectual property has become, and if you have an asset that is very valuable, you'll do what you can to seek to protect it and extend that protection even when it backfires, as I believe it has in the medical area with the HIV/AIDS pandemic and the price of hep-C drugs.

I am of two minds when I think about what the private actors were doing in the trade negotiations, negotiating for changes to the World Trade Organization to better protect intellectual property and post-WTO efforts to protect digital content. Because the Internet was barely up and running when the Agreement on Trade-Related Intellectual Property Rights was concluded in 1994, the rules didn't really apply very much to the digital realm. That's been a big challenge.

But the actors in [*Private Power, Public Law*] present themselves as the cutting edge of the world economy, and that they're the leaders of a bold new successful economic enterprise based on the ownership of intellectual property. However, when I look at some of these industries, whether it's Hollywood or the pharmaceutical industry, it looks much more like the use of law and regulation to increase rent-seeking to hang on to the last lucrative years of business models that are being deeply challenged by technological revolution. We see that really clearly in the case of newspapers, that the digital age has really hurt print. We have seen it in the entertainment industry, with movies and music, where people are used to downloading things without paying for them—at least the younger generation is—and there's a real divide between the people that argue, information must be free, and those who want to lock up property rights and control or ration access to the fruits of those rights.

Blayne: Is it an overstatement to say that intellectual property is becoming more important in influencing technological development?

Susan: Oh no, absolutely I think that it's becoming more important in the global economy. Nanotechnology, biologic drugs, I mean all of these cutting-edge technologies. People want to be the first movers in those areas and get advantages for innovations in those domains, and I think there is also an important role that technology has to play in solving a lot of global challenges and that's been clearly recognized, especially in the case of climate change, shift to more sustainable technologies for generating power.

Blayne: Has international studies kept up with the importance of intellectual property?

Susan: Well, I certainly think that it would be nice if more scholars were looking in this area. I do believe it is really important, but I think it's incumbent upon us to make the case that it is really important and that's why they need to pay attention to it. Maybe we're not doing a good enough job about getting the word out.

I also find that it can be a little intimidating to jump into this area. When I first started working on these topics back in the 80s I hardly found any political science or any of the International Relations literature that really addressed the issues that I was interested in looking at. And so, I ended up having to read a lot of law journals, a lot of economics journals, even some business journals for where people were talking about it, and having these kinds of conversations. And history as well. I found a lot of great stuff, but very little of it was found in political science. I think that's because there were very few people at that time writing about development. There had been some good work done on multi-national corporations by Stephen Hymer, and Robert Gilpin, of course, had an important book back in the day. But there wasn't much. Stephen Krasner had a book about the North–South divide, the politics of that. But there was really very little out there.

So, it can seem [that] getting into these issues can seem rather technical, rather difficult, rather arcane. And you do have to master a certain amount of technical detail in order to write knowledgably about it. And you have to be pretty interested in it, I think, to follow it down. And I think it's our job, if we think it's important, to try to translate it in a way people will see the absolutely central political nature of these things.

I mean, property rights, there's nothing more political that I can even imagine, because there are winners and losers in all of these things. There are winners and losers in debates over safe harbours for internet service providers, there are winners and losers when we have the NSA being able to spy on us without our consent. These are, at heart, very important political issues, and I think that's what makes them legitimate objects of study for political science and for international affairs.

Blayne: Do [the core IR theories] have any purchase when we're talking about technology? Or, how does technology influence these concepts?

Susan: Security is not my core object of study, but I think that new technology has a double-edged sword quality, especially in cyberspace. There are extremely powerful cyber-tools that governments can use, like Stuxnet, we used some malware to slow down Iran's nuclear capability development, but there are also very big vulnerabilities, which explain the rise of cyber-security as a big industry now. So, in security, it's a double-edged sword, but I think people that write about security have talked about capabilities and vulnerabilities forever [laughs], whatever kind of technologies are involved. I don't think that's new, thinking about that double-edged sword. Nuclear weapons, that was a disruptive technology, right? Thinking about these things, the way that they have been thought about in the past is still quite helpful.

And the question of power, we can think of technology as being a new—it's not new, technology is not new, but the digital realm and some of the new innovations like biologic medicine, as they can be something that will help support or promote your power, they can be strategic assets in that sense. They can be strategic in terms of generating revenue, making your economy grow, increasing your capabilities, whether it's curing disease or competing with other firms, so power still is useful. Technology can be thought of as another arrow in your quiver. Basically, it's something that could help bolster your capabilities.

In terms of global governance and global order, I think we're still trying to understand the role of the digital communication and information technology revolution on global governance. In some ways it's made things easier, in terms of reducing the cost of collective action, but in other ways it's made it somewhat confusing. And I think there have been some interesting studies about what they call slacktivism. You remember the Joseph Kony 2014 campaign, which was based on things that perhaps weren't true. People being flooded with so much information that they're not getting the full context or the analysis.

But, again, all these things can be used to promote better governance and accountability and whistleblowing, but they can also be used in way that can be quite opaque, as we see with the Panama leaks that the ability to transfer money at the touch of a button has made it a lot easier for people to hide their wealth and move it around, and the technology had facilitated that. So, I guess all these things are, I guess, double-edged swords, but I think technology can be understood in terms of being a security asset or vulnerability, a power asset or vulnerability, and then a governance asset or vulnerability.

Blayne: Is there anyone who's doing a particularly good job at studying these issues right now?

Susan: I think in International Political Economy, I would say the scholars that are working on regulation are doing a better job of it, [those] who look at business regulation because that's where a lot of the action is right now. You know, how are these technologies being regulated and what are the implications of addressing them with this law or that protocol or another one? I think that came about in a very interesting way in the fight over the *Stop Online Piracy Act* and the *SOPA/PIPA* in January of 2012 when Internet experts argued successfully that, in fact, if you adopted the policy the way that it was written, you're actually going to dramatically reduce the security and stability of the Internet itself. These things are really consequential, but I think looking at the way that it's being regulated leads you quicker into those questions than perhaps other ways of thinking about it.

Blayne: In your 2013 article, "Revenge of the Nerds," you argue that the SOPA/PIPA fight is that bringing digital technology together with these social movements could potentially form a counter-hegemonic movement. A lot has happened since then, but is there still the potential for a counter-hegemonic movement in this area or generally?

Susan: I think there is, and depending on the day, I'm either more hopeful or more pessimistic about that. But, I guess I'm old-fashioned in that I believe in the old enlightenment perspective that knowledge is power, and the more that people can be informed about what's going on and how it affects them, it creates opportunities for political agency that with these technologies, just a few people can make a lot of important things happen. If you think about [Jody Williams], who started the ban-the-landmines campaign, basically on her laptop in Vermont. She was able to start something quite important, in my view. And I'm really interested in people like Edward Snowden, Chelsea Manning, Julian Assange, and now whoever leaked those documents on the Panama shell companies. I think there's potential for people to make a difference and different kind of global governors to exercise agency and have an impact in way that can be quite important.

So, I find that information can be a very powerful tool to spark mobilization and spark agency on the part of people who want to make a difference. I mean, it's not a level playing field, but I think it has opened up opportunities that make it easier to organize and easier to publicize what's going on. I think a good example are these cameras—that's again a double-edged sword—but all these cameras showing police officers gunning down unarmed black men. The whole Black Lives Matter movement, I think has gained a lot of traction because of technology.

Blayne: Who are some other the thinkers doing applied work that you look to? Who are your go-to theorists?

Susan: I'd have to probably start with Polanyi as a go-to person and absolutely Susan Strange. I should have mentioned her before because her work and Ernst Haas' work on knowledge and power and epistemic communities is very important. But Susan Strange's work is the work in political economy that really spoke to me when I was in graduate school. She was someone who offered pretty powerful critiques of mainstream American International Relations, and I could relate to her critiques in the sense that many of the off-the-shelf International Relations theorizing at the time I was in graduate school didn't apply to negotiations and relationships between highly unequal players. Most of the political economy literature was derived from studies of the OECD countries, and you could assume that they had a lot of interests in common and that power relationships were relatively symmetrical. But I was interested in the global South and, you know, just because they signed a treaty doesn't mean it's good for them and that everybody's happy. I think that International Political Economy underplayed a lot of coercion, a lot of asymmetry and it created this sort of happy-jolly world where people want to reduce transactions costs, and I felt that they were really obscuring some of the really fundamentally political questions.

It wasn't just about designing institutions; it's about the process, it's about what the substance of those institutions is, and who wins and who loses. And Susan Strange always had at the center of her work: this is good for whom? I started my career in political science as a student anyway studying political theory, and we always started with the normative question: how people ought to live. Her work resonated with me very strongly; she was one of the few. And I also, when I was starting out, I read Craig Murphy's work and Thomas J. Biersteker's work and they engaged some of these issues in a way that a lot of other scholars weren't. Peter Drahos' work of course was very, very important to me. And the work of the new legal realists right now, like Greg Shaffer, Heinz Klug; there are a lot of interesting—Terence Halliday's work on bankruptcy—there are a lot of really inspiring people working out there that I go to. Margaret Archer's work, she's a sociologist, I found her work also very useful in trying to square the circle between structure and agency.

Blayne: Final question: if you were setting the agenda for 'Technology and International Relations', what are some of the questions that we need to answer? What would you suggest to research students that they look into because it's important?

Susan: Some of the questions I think we need to answer are: To what extent can states control technology? To what extent can states use it? I think that's a really interesting question. What role does private governance in standard settings play in the development and deployment of technology? What are some of the risks associated with particular technologies? What's different now versus in the past? And for that you have to look at some histories of technology.

What are the opportunities, what are the potentials, what kind of problems might technology be able to solve? What dangers do some particular technologies pose? Who and what are these technologies good for and who do they tend to harm? I think that's a question like Susan Strange's good for whom? I think that's one that runs—or should run through—anyone's consideration of issues having to do with science and technology.

Blayne: And it also has that balance between, as she put it, authority and markets as well.

Susan: Yes, yes, and that's why I go back to Polanyi about the relationship between states and markets.

Blayne: Thank you very much for agreeing to this, this was an absolutely fascinating conversation.

Susan: Well, thank-you, Blayne. I just hope I actually answered your questions [laughs].

Blayne: Well, I learned something. That's got to count for something.

Susan: I'll say that's a good thing [laughs].

WORKS REFERENCED

Halbert, D. (2016). Intellectual property theft and national security: Agendas and assumptions. *The Information Society, 32*(4), 256–268.

Monga, V. (2016, March 21). Accounting's 21st century challenge: How to value intangible assets. *Wall Street Journal.* https://www.wsj.com/articles/accountings-21st-century-challenge-how-to-value-intangible-assets-1458605126.

Ocean Tomo. (2015, March 5). *Annual study of intangible asset market value.* www.oceantomo.com/2015/03/04/2015-intangible-asset-market-value-study.

Sell, S. K. (2003). *Private power, public law: The globalization of intellectual property rights.* Cambridge: Cambridge University Press.

Sell, S. K. (2013). Revenge of the 'Nerds': Collective action against intellectual property maximalism in the global information age. *International Studies Review, 15*(1), 67–85.

CHAPTER 4

Everyday Tech: In Search of Mundane Tactics

Michele Acuto in conversation with Saskia Sassen

Abstract In this chapter Saskia Sassen, discusses tales of how technology (re)articulates political-economic relations but also requires a 'mundane' sensibility. In this chapter she does so building on personal memory, interdisciplinary view of IR and curious facts, highlighting serious matters of inequality and power. Engaged here in conversation with Michele Acuto, Sassen calls for empirical engagement 'on the ground' and continuous conceptual (r)evolution and self-examination. In particular, building on her scholarship on markets, cities and spatial appropriation, she offers as reminder that, when it comes to the role of technology in international relations theory and practice, no example is too mundane to matter.

Keywords Mundane · Globalization · Everyday technologies · International political economy · Scale · City

M. Acuto (✉)
University of Melbourne, Melbourne, Australia

S. Sassen
Columbia University, New York City, NY, USA

35
C. Kaltofen et al. (eds.), *Technologies of International Relations*,
https://doi.org/10.1007/978-3-319-97418-7_4

"Whatever, no I don't mind at all where we start... often times I just start!" It is hard to capture Saskia Sassen's restless train of thought as we jump between separatist struggles and nanny apps, global economic downturns and the steam engine. Sipping a glass of Chilean white in her sunny central London flat, packed with multilingual editions of her works and imposing musical instruments of her historian husband Richard Sennett, Sassen tells us tales of how technology (re)articulates political-economic relations but also requires a 'mundane' sensibility. She does so in her unique style between a laugh and a statistic, a personal memory and a curious fact, never skirting away from serious matters of inequality and power.

Sassen is no doubt one of the most recognizable and widely cited contemporary social scientists. To be certain, Sassen's popularity bridges well beyond the walls of academia as one of her most famed research themes, the "global city", has rapidly proliferated amidst urban practitioners the world over becoming a staple of city politics throughout the 1990s and 2000s. Nonetheless it is no easy task to label Sassen with a specific disciplinary affiliation. Her bibliography, now impressively counting in the hundreds and with translations in over fifteen languages, is a constellation of musings on the dynamics of globalization, transnational flows and urban transformations, cutting across geography, sociology, political economy and urban studies, to name but a few. In this sense, Sassen's influence in International Relations theory, especially when it comes to thinking of globalizing processes and their socio-economic consequences, is variegated and not easily captured organically. Yet, connecting the themes of globalization, political economy and societal transformation, Sassen's work is ultimately a call for empirical engagement 'on the ground', continuous conceptual (r)evolution and self-examination—all themes that echo in the interview below. When it comes to the role of technology in international relations, both theory and practice, Sassen reminds us no example is too mundane to matter.

> Michele: Great, let's make a start. If you don't mind starting from the obvious then, what are 'technologies' for you?
>
> Saskia: Technology, if you take the term seriously, has been there forever—really an enduring feature of humanity. This is an obvious thought but also key reality for politics because, when we talk about technology today, the tendency is to go for the more advanced sectors

and I think that is in fact a weakness in public and academic speak. We should recognise that technology comes in many different shapes and we also need to think of the more mundane and less popular technologies. We shouldn't be seduced by the cult of the new and remember the truly revolutionary emerges in many shapes and places. To give you an example of that, when I was in Nairobi in 1985 for the 3rd UN World Conference on Women, I met the Women of the Polisario[1]—amazing warriors showing all kinds of technological wit—working in one of the least hospitable places in the world like the Western Sahara Desert. They had come up with stove technologies that allowed them to do what otherwise could not do by capturing solar power for very simple needs. That stove became an important instrument that travelled, partly because of the Nairobi conference, across the world both as innovation as much as symbol of the Western Sahara struggle, but the invention mattered less than the power of finding ingenious solutions to pressing and very localised needs. And it came from Women in a very male-dominated society. There are countless examples like this, but rarely we think of these when we say "technology" and people's minds jump straight to the Web. Or an app. Actually, just certain apps, not the really mundane ones. Remember to tell you of the 'nanny app'! So whenever people reduce the meaning of technology to the latest 'tech' I'm always reminded of those cases, and no doubt there are many such moments that one could think about and would take us out of the latest innovation into the role of technology as constituent in civilization. After all we all know that technology's rate of obsolescence is increasing rapidly, so what really matters is the socio-technical assemblages constituted around certain technologies, and the logics embedded in them, much more of the piece of tech itself.

Michele: *Taking this into account, what role do technologies play in your empirical work, and how do they influence your analytical perspective?*

Saskia: Oh I have had a long affair with technology! I got involved in that more than 30 years ago now, when the 'world wide web' mostly did not exist, and the meaning of communication technology was set on this very standardized world, which presented the dominant way of 'technology', and when you looked at emerging finance you could see a whole suite of engineers and scientists beginning to work a different type of instrument from that which existed. Yet back then "ICTs" like those deployed by AT&T and the likes of these communications giants were effectively monopolizing the meaning of technology, perhaps making it narrower than it should. When the world wide web comes into this picture, then, you have yet another narrowing of the meaning of "tech" from that of the "ICTs". And this presents a dissonance from my own experience where I try and engage with technologies more broadly. So when I originally got engaged in technology I had a series of experiences of what 'innovative' technology meant, one of which was that of the women of the Polisario, which still encourage me to look at more than the tech itself... into its logics and assemblages one could say.

You see, and I've argued this with sociologists and geographers and even international studies, many social scientists see technology as the driver of fundamental social transformations. This is not necessarily wrong, but I we the get too fixated on the technical properties: we construct the relation to the social world as one of applications and impacts of technology, often forgetting the logics embedded in technology, the people using them and the assemblages around them. The challenge then, as I've written many times, is not so much to deny the 'weight of technology', but develop tactics that allow us to capture the complex imbrications of technology and society.

For myself technology is also very much of an instrumental entry point: I certainly have learnt a lot about technical aspects over the past decades but I'm no expert, not an MIT engineer... When I began to work on the digital from the perspective of finance, MIT was very much the leader on this front. So I got hooked quite a bit into that financial moment, but I really want to emphasise that what is critical is to appreciate the mix of all of these actors and innovations, from AT&T to the women of the Polisario. Altogether they constitute a thicker multi-actor world of technology: in this diversity, at least to me, the term 'technology' was an extraordinarily ambiguous word... without having even begun to mention that whole reality of how we build and make the world, including engineering and much more. What these stories, from the desert to high finance, tells me about technology is its place-boundedness and the 'thick' multi-actor social environments within which technology operates and where political-economic assemblages are constituted.

Michele: Interesting: do you think that that technology is necessarily embedded into place or is there a more-than-local dimension to these dynamics?

Saskia: In a way, I think it is necessarily local and that is why it has specific attributes. But this doesn't mean rejecting the global (and I don't like speaking of 'it' as a 'thing'—but global as adjective, as process...), but rather that the global is necessarily articulated through the local, it's "embedded" in the way that I for instance speak of "embedded bordering" now as cross-border geographies constituted through the specificity of each place, country, people.

Which takes me to the 'emergency nanny'! It's a great example because of this embedded-ness when we speak of technologies and immediately turn to the latest Android apps, we tend to focus on those for the tech savvy and the professionals. You do that too when you talk of your smart cities, but the emergency nanny is great reminder of the mundane life of these technologies. The emergency nanny is an app that in my reading has a first step addressing a need based demand but has a deeper meaning in that it mobilises neighbourhoods to emerge as a social support system.

Let me explain: low-wage working parents that need to be in contact with their children often operate within closed and very unflexible {sic} work systems, so the app helps them connect with other adults who are stable or static in the neighbourhood to act as their 'emergency nanny' if work schedules change. As a low wage worker, you can very discreetly engage in a social network that offers a solution to a very important but very mundane crisis that in all effect endangers the very texture of society. The first step is solving a problem of course, but the second is that those people begin to create a cooperative platform. The app prevents the tension of working parents between being available for their children and following work rules. Little networks like this are being created everywhere, beyond the big and well reported digital innovations, that are effectively platforms of solidarity and, or, of trust that can be deployed when the social texture is at threat. These are the sort of "uninteresting" and mundane interventions of that digital world which should be better appreciated and not just criticised. The trajectories of collaboration that they create, like the emergency nanny's trajectory for the mobilisation of 'neighbourhood', are really fundamental in today's society.

So this might not be a 'global' movement, but cross-border geographies of technology are at play here and you know that many different localised communities can have a similar 'mode', and what it does is that it builds a network of trust within low income contexts, and neighbourhoods—in a sense still valuing the local in a de-territorialized digital global. A little seed of trust in a complex world. I like the idea of embedded seeds: it speaks to a multiplication of the local which becomes a global event. This is not the cosmopolitan route to the global: this is about the global as a multiplication of the local, and these are types of sociability embedded in people's mundane actions and activities.

Michele: If you had to build on this, what are the most significant challenges and opportunities that technologies raise for world politics today?

Saskia: Well, what is the steam engine of our epoch? Perhaps it is a rhetorical question, as many people would say "its digital technology"—a typical answer really. But I don't agree with that: I think this was the case 30 years ago, but today the digital is, you might say, infrastructure—it is necessary but indeterminate. So the actual question should be "how is it getting used?" Digital technology is fundamentally infrastructure in our time, not a 'steam engine' but the brick and steel of much socio-technical interaction. To me the core innovator today, the one that is the steam engine today as opposed to 30 years ago, is high finance, which has altered all kinds of spaces, dynamics and sectors.

The steam engine didn't change everything—it created a new logic, and the same is with high finance today. If you were able to fly on a little plane over England in the 1800s you would have seen mostly rural space and farmlands. But in reality that sheep on top of that green hill was no longer rural: it was part of a whole new circuit that connected it to the factories in the cities. You couldn't quite see it from far above but it was viscerally changing socio-economic relations. I use this example in *Expulsions* to illustrate a key feature of this 'steam engine' role of technology—a lack of transparency that attaches itself to the material. That sheep has been lost, no longer belonging fully to the farmer, but part of a foundational transformation of that time, powered by the steam engine. And of course, as the steam engine's disruptive power did not last forever, so IT 30 years ago versus its infrastructure rile today, and so it will be with finance. In this sense, I argue that {high} finance is like that because it has succeeded as a steam engine of our time in foundationally repositioning the material of our epoch, and secondly in inventing instruments that allowed it to de-border and re-border radically founding institutions of our time like national economies and national laws.

Michele: *A propos of this, you've suggested that "Periods of rapid technological transition have heuristic potential"* (Sassen, 2008); *Can you explain what this means for IR?*

Saskia: Yes as I've said a few times this is because these moments are a window into a insides of technology unveiling the logics and material formations they are embedded in. And this of course opens up all sorts of political questions IR students must unpack. In particular, their potential is that they promote the idea of disassembling the socio-technical assemblages that uphold, and this is a heuristic potential because whilst not necessarily calling perfect methods, they offer a sufficient window into the complexity of how power is made and embedded. It's not always a job for refined theory though...

Michele: *So still regard yourself as a 'carpenter*[2]*?*

Saskia: Yes of course! I like to dismantle and rebuild. That is a common tactic in many things I have researched, which lets you capture the complexity of technology in society by going beyond function and disaggregating things: you pull them apart, and understand embedded logics. That teaches you about technology too and not taking it just for its form or function. For instance, the 'global financial crisis' as we call it is not a crisis of financial or technical matters but a crisis of confidence. In the late 1990s we had actual financial collapses when counties like Brazil, Argentina or Korea went broke, but in 2006 instead doubt enters into the picture. It is a moment of international ramification of crisis hedge funds (the main buyers land in the global south) that confront a crisis of confidence: it is the consequence of a financial project that starts brilliantly but then power abuses technology. By 2007, as foreclosures and bankruptcies skyrocket, and the instrument of the financial steam engine keeps on growing, and a key moment is the its sudden visibility to the naked eye. It opens up the technology at play in a very real sense as people relate to houses and protests and displacement. This crisis is one extreme condition that makes visible the innards of the system, against the *invisibility* of the massive expulsion of people: technological intervention that is deep and wide across people. Yet we still struggle to see the extent of 14.5 million households being emptied, so what really engages the theorist in me is the invisibility of this material displacement: when confronting technologies we must talk bodies and materialities not just theoretical elements. We are talking real people with tangible and relatable needs. The role of the scholar, or at least some scholar (… *the* scholar is such a heavy term!), is not just to critique or describe: with a need to also bring in a normative argument to the enable a deployment of technology that is for people and rights, and that can start from the mundane too.

Michele: So what should we focus on when studying technology in IR?

Saskia: What to focus on? Well, the *what* is one thing, as the actual content. The 'what' is always constituted in so many very different often very small but certainly very valuable instantiations. Perhaps one way of thinking about research agendas in IR is firstly to recognize there is always a much larger operational space that needs to be seen and problematized beyond what is commonly argued. And secondly that the political gets made, constituted, thought about and argued about in many domains where we do not think politics. Yet in many of these, whether it is the Polisario Front or the low-wage workers in major global cities, people fight for technological advancements and technical revolutions, and these mundane, less seen, reality also deserve a serious look.

So when we come to the issue of technology, I think that the fascination with the latest development, the latest discovery, the latest fantastic gadget, is the ruling mode for many international theorists and thinkers. I find that very problematic if not actually extremely unacceptable equally by practice and theory. We should not shy away from perhaps more mundane in their own way questions of politics we also often discard in favor of the new—in an academic sense too. Along this vector, for instance, we could and should ask: where does government governs? Some governments assume the capability that they are governing other parts of the world, with the USA as a prime candidate. But where is the governing actually happening and how? And what is the relation between governing and governed... or better what is the relation between power and powerlessness?

This is not to say that digital technology doesn't have a place today. Quite the contrary, we now see how simple technology is in reshaping our lives. But as we see, in practice or in theory, we should also make sure that we don't lose people and mundanity, and that we don't stop at the functionality of the 'tech'—taking its logics and its assemblages into account. And this can tell us much about the very shape of politics, international or other, in the present age.

Powerlessness is not simply an absolute condition that can be flattened into the absence of power. Under certain conditions, unpacking certain technologies shows us how powerlessness can become "complex"... by which I mean that it contains the possibility of making and remaking the political—something that we can see in global politics but perhaps we should also, as I've written about, recognise on the 'global' street where all of the cross-border processes hit the ground. One way of thinking this in IR, but I am always afraid that the reader will misrepresent it as it is such a 'contaminated' term, is to understand there is so much *making* happening in these more mundane examples that deserves international theory.

It is also important to place these discussions in a much wider operational, embedded, and material space that has a real mundane flavour: think about the huge waste implication of the digital revolution, the masses of electronic waste and the global logistics required to keep it all together and 'new'. But think about the everyday challenges for low-wage workers to keep their neighborhoods functioning too. Much of the digital debate and its discussion often miss it. And we miss these are all elements of a bigger picture which we can capture, in an IR sense, but need not to forget its embedded realities. I can understand this inherent bias: it's just stuff of less interest. We become passionate about what engages us fully and less passionate about the routine and the mundane, and this is very much an outcome of the way we function in the social system today.

Michele: Well that's a good way to wrap it up, from the street up... 45 min and six thousand words, I'm afraid we'll have to do some cutting!

Saskia: Oh that much? We were really just getting started... More wine?

Notes

1. The Polisario Front is a Sahrawi rebel national liberation movement aiming to end Moroccan presence in the Western Sahara, who has had a well-recognized 'front' of women fighters. See Loveday Morris. Women on frontline in struggle for Western Sahara. *The Guardian*, July 16, 2013. Available at https://www.theguardian.com/world/2013/jul/16/women-western-sahara-independence-morroco.
2. We refer here to Sassen having described herself as a 'carpenter' of social theory in response to Aiwha Ong's self-description as an anthropologist-*bricoleur* in their joint interview on assemblage theory (Sassen and Ong, 2014).

Works Referenced

Sassen, S. (2002). Towards a sociology of information technology. *Current Sociology, 50*(3), 365–388.

Sassen, S. (2008). Re-assembling the urban. *Urban Geography, 29*(2), 113–126.

Sassen, S. (2017). Embedded borderings: Making new geographies of centrality. *Territory, Politics, Governance* [online first], pp. 1–11.

Sassen, S., & Ong, A. (2014). The carpenter and the bricoleur. In M. Acuto & S. Curtis (Eds.), *Reassembling international theory* (pp. 17–24). Basingstoke: Palgrave Macmillan.

CHAPTER 5

Curiosity, Criticality and Materiality

Can E. Mutlu in conversation with Mark B. Salter

Abstract In this chapter Mark B. Salter, current editor of *Security Dialogue*, discusses with Can E. Mutlu the meaning and significance of technology for International Relations in light of his eclectic work. Salter, perhaps best-known for his dynamic presentations and engaging intellectual approach and recently for the two-volume project *Making Things International,* traces his engagement with technology across a vast field of contributions ranging from civilization in international politics, the genealogy of the modern passport, and critical security studies, touching on Foucauldian and Bourdieusian notions. In the conversation, Salter reflects on the recent material turn in IR and the expansion of this as a significant research area within critical consciousness in IR, with more and more people working on materiality, science and technology studies, actor-network theory. He points to the importance of remembering that we are not the first generation to experience this kind of epochal change, and that emancipatory change happens through engagement, and how technology is shaping the encounter with the

C. E. Mutlu (✉)
Acadia University, Wolfville, Canada

M. B. Salter
University of Ottawa, Ottawa, Canada

C. Kaltofen et al. (eds.), *Technologies of International Relations,*
https://doi.org/10.1007/978-3-319-97418-7_5

45

Other—reminding us that scholarship can and should start with curiosity and intuition.

Keywords Actor network theory · Security · Borders · New materialism · Foucault

Sitting in a conference room at the University of Ottawa's Faculty of Social Sciences building, overlooking the Rideau Canal in Ottawa, Canada, it is hard not to think about the role of this significant technology—i.e. river locks, on the history of Canada, on the fate of its native and settler populations, the selection of Ottawa as a national capital, and even the history of British Empire. A very relevant and interesting backdrop to a conversion about the meaning and significance of technology for International Relations with Mark B. Salter, Professor of International Relations at the University of Ottawa's School of Political Studies. Professor Salter is the current editor of *Security Dialogue*. He is perhaps, best-known for his dynamic presentations and engaging intellectual approach. He is also the editor of *Making Things International, Volume I: Circuits and Motion* (2015), and *Volume II: Catalysts and Reactions* (2016). His contributions to IR cover a vast field, ranging from a noted manuscript on barbarians and civilization in international politics (2002), to a genealogy of modern passport (2003), as well as a number of edited volumes, including the one we edited together on *Critical Methods in Security Studies: An Introduction* (2013). His most recent contributions to IR, Materiality, and Technology came in the form of *Making Things International*, volumes 1 and 2.

> *Can*: *On reflection, how do you feel technology has been present in your work? What kind of an influence has it had on your thinking about International Relations (IR)?*
>
> *Mark*: Thank you for the invitation to be interviewed, I am really happy to speak to you, and have this chance to reflect on these questions. I am going to answer in two parts. One, is that when I started my undergraduate, the library was still paper-based. You still went to a card-catalogue, and a significant portion of research by discovery was going and seeing what the librarians had catalogued next to books that you were interested in. So, for me, my research has always been curiosity-driven.

The haphazard approach of going to the "JZ" call numbers and seeing what was new, that was the way I stumbled into lots of interesting theories about international relations. The evolution, or progress of my career, has been coincident with the digitalization of research. I went to the public records office to do work on the passport, with paper. There was a digital catalog, but then it was going through boxes of files, of departmental reports that really, illuminated for me some important stories about the evolution of the passport. The way technology has changed not changed my curiosity but changes the way path curiosity takes. Because things are connected in a much flatter way, through the Internet, through the digital catalog, that means that I am much more likely to stumble upon anthropological works, or works in cultural studies then I was before, when I was stuck to the JZs, or when I was going through the boxes of documents at the passport office. So that's kind of one way in which technology has really altered the way that the path of my career. I also have a hard time imagining how I would do some of the editorial work without the Internet, without the digital connection and the digital platforms that I work on. The second way I want to answer that question is by saying that my first postdoctoral project on the passport was my way of trying to understand the connections between identity, the state, and power through what I understood the passport to be in the larger Foucauldian sense of the word. I think that I had experienced myself as an immigrant, as someone who went through a nationalization process in their early teens, who moved around a lot. I was very conscious of the connection between identity and citizenship at a very young age. That really formed a core sense of my sense of the world, my sense of politics, my sense of myself. The moment I read Foucault, who gave a larger definition of technology, and that it was not just about an object, but a way of seeing the world a set of tools and instruments, that really clarified how I had experienced those important questions of identity, citizenship, being, belonging, and power. The moment that I had that version of technology, that really helped me develop a productive framework for my research so that I reconcile the soft idea of identity and the hard idea of the passport in a really useful way.

Can: Thank you. How have approaches to the relationship between technology and concepts like power, security and global order develop in IR. I guess? you touched on this a bit in the second part of your previous answer, but it may be great to flush out these ideas further given the nature of your work and such relationalities it embodies.

Mark: I think that is a really interesting question. It seems to me that that IR went through a phase in the mid-to-late 1980s to the early 2000s that took a theoretical discursive-turn that moved away from technologies. If you go back, however, to early post-WW2 IR, you see a great concern about technology, about science, about nuclear technology, about raw materials, about the geopolitics of natural resources, the geopolitics of oil. There is a real concern with technology, "there." I think that the critical-turn and the move towards structural realism, the move towards neoliberalism and neorealism, that really moved IR's attention away from the material and the technological towards the discursive and the abstract—I say that not as a critique but merely as observation. The way that IR came back to the technological and the material was both through feminist works like Donna Harraway and Cynthia Weber working on visuality, as well as others working in that critical tradition, working in that wider sense on Foucault, but also came back through an engagement with that mid-to-late stage Foucault that was not simply about discourse but rather about institutions in the broadest possible sense that included government operations, objects, discourses, expertise in that way. I think I am indebted to you, for really directing me to the new materialism. When we started the project that eventually became the *Critical Methods in Security Studies: An Introduction* in 2009, we mapped out the big "turns" in critical security studies, we talked about the ethnographic-turn, the discursive-turn, the Bourdieusian practice-turn, and the affective-turn, using affect, which was becoming more prominent in social theory, it was you who brought up the new materialist turn that is borrowing from Science and Technology Studies. My experience, or my recollection was that in 2009 we struggled to find ten or so people who did that kind of work. It was in the air, and people were starting to read it, but research projects had not gone their full lifespan so that there was not a lot of published work using either Latour, Mol, Law, Bennett or Connolly, or others. At the time, Connolly had already written *Neuropolitics* but had not yet published *The Fragility of Things*, Jane Bennett had written articles that had the ideas presented in *Vibrant Matter* but had not published the manuscript, so at that moment, the awareness of technology, or that way of understanding technology, in particular the science and technology studies, actor-network theory was starting to emerge, but had not yet consolidated in IR. Within those two/three years, between 2009 and 2012, that we put together the section on material-turn with 5 great contributions in the edited volume, that research area became a significant emerging field within critical consciousness in IR. There were a lot more people working on materiality, science and technology studies, actor-network theory in IR in 2012, then there were in 2009.

Can: *I think those approaches to materiality, science and technology studies, and new materialism in particular that we covered in the chapter, definitely provided a theoretical framework for a lot of people's curiosities working on drones, computers, networks.*

Mark: I think that as we agreed all those years ago, IR is a piratical discipline. This was true with European scholars, but also with critical scholars in North America, that Science and Technology Studies was a way to connect the Foucauldian concern with discourse and the Bourdieusian sense of practice with those technological aspects of "things" like drones, algorithms, surveillance, that all of a sudden had become really visible to the critical consciousness.

Can: *I want to now touch on something that you mentioned at the beginning of your second answer. To what extend, can our ideas about the role of and implications of industrial age technology help us understand the challenges and opportunities of the information age. I guess this sort of touches on your reflections on the post-WW2 interest in technology in IR, and perhaps going even further back and reflect on a broader understanding of technology that pre-dates WW2, and how that can inform our current understanding of contemporary technology?*

Mark: I think it is important to remember that we are not the first generation, or first era to experience this kind of epochal change. When I look at the history of the passport, when I went through those boxes at the public records office, you could really see the way the government was struggling to manage mobility in a radically different way, or struggling to manage a refugee crisis after the first world war, that seemed completely unparalleled in their experience of human history. I think that one of the conceits of the information age is that this is a revolution of unexpected or unimaginable proportions. I don't want to dismiss that, in the sense that it is absolutely correct that we are in an Anthropocene age where the industrial, and the post-industrial way of relating to the ecosphere has made irreparable changes in ways that we could not predict or do not yet fully understand and may not be able affect. However, this is not first time human consciousness or human societies have struggled with radical change. If we take seriously the debates about the nuclear age, those were about connectivity, about complexity, about the capacity of humans to make decisions that have apocalyptic effects. If we look at the nineteenth century, and the emerge of the steamship, or the emergence of the telegraph, we see similar kinds of epochal change. For me, the most important lesson is to be to able to have a bit of distance, and say how have societies, states, humanity reacted in these eras of fundamental epochal change, rather than chase ahead and we need a completely new theory or concept to understand this new age.

Can: So if I get it right, what you are saying is that historical reflection is necessary to understand transformations in a continuous way to be able to say that any transformation is not just exclusive to this period or that period, but rather it is part of a historical trend?

Mark: I hope this does not sound philosophically conservative, as I do not mean it to be, but while we may need new theories and ways of understanding the empirical changes to the world, the human reactions to those changes are not dramatically different in the 21st, as they were in the nineteenth century. When you read Conrad's *Secret Agent* about an anarchist terrorist threat and the fundamental way that sort of *undoes* basic notions of human society and trust, that seems really similar to me to the way we are talking now about extremism and terrorism in the West. That same sense of completely new ideology that involves violence, that involves a radical destabilization of the status quo, that maybe new to some of us in our experience, but that is not a new experience in society. While the technologies may be revolutionary and the changes themselves are novel, we can look to the way societies have reacted to those changes in the past in ways that can be really useful to understand how they are reacting now.

Can: My final question, once again comes in relation to your reflections. What do you see to be most significant challenges and opportunities for international relations in the context of technological change? Less so than scholarly approaches to International Relations, here I am interested in getting your ideas on the role of technology on everyday practices of international relations.

Mark: One of the articles that I read on a bus was Barber's "Jihad vs. McWorld" (1992) back when it came out that made me an IR scholar, was the key insight he had in that article that: the empirical forces of globalization, the revolution in communication and technology that all of a sudden billions of people in contact with one another, the rapid expansion of this particular form of capitalism, and the shifting modes of governance, that these new eras brought about, those three radical changes could have two dialectical opposing results in societies. That some societies would rush to a global world, and others rush away from that global world with equal vehemence and speed. That really indicated to me how in each kind of epoch, in each kind of revolution, there is equally the opportunity for progression or conservatism. That there was nothing inherent in the technology of the cellphone that made it a force good or force for evil. Or, as I'd rather, if we chose non-moral concepts, there is nothing inherently integrationist, or separationist about the cellphone. I think that is great example. The cellphone has empowered an entirely new way of banking in a country like Kenya, where cellphone credits has allowed for a whole set of people to be bankable and allowed for a whole different kind of economy to emerge. Cellphones also is a crucial part of a technology that renders occupying forces in places like Iraq and Afghanistan vulnerable to improvised explosive devices. There is no way to make some kind of prediction to say that there is something inherent in this technology that is emancipatory or conservative. However, I think that, at my core I am an idealist in the sense that I think the way towards emancipatory change is through engagement. It is through encounters with the other. One of the things that I see flattened communication technologies are doing, one of the things that flattened economics is doing, or flattened governance is doing—however problematic all of those dynamics can be—is by allowing new political coalitions to form, that are outside, or parallel to state sovereignty, which have allowed for so much institutionalization of human misery and inequality. Even though I do not find an inherently emancipatory potential in something called "the People," I think that the more avenues there are for new connections to arise, the more resilient the human society becomes, and the greater potential for interaction. I think of my own experience at the American University in Cairo, before I arrived the production of knowledge from Cairo was dependent on the typewriter, and the postal service, and the telephone, and maybe the fax machine. That just meant that there were certain structural barriers to the production and the dissemination of knowledge; it was not just how expensive it was to be in Cairo, or to get out of Cairo to go to international conferences, but to send out an article was unreliable, and random and would require a lot of work and effort. Whereas now, I communicate with scholars from Egypt and the Middle East all the time through email that skirts around not just the state authority but also the institutions. I see that as having a real potential.

Following from my earlier comment on how Foucauldian notion of technology have allowed me to really incorporate elements of my experience and my institutions about power and state controlled to be reconciled, it is really important to be open to that broad sense of technology. Technology is not just algorithms, the machines, and objects, but way of seeing and understanding the world as a set of *operants* available for utility. It is important for us to be mindful that technology is not a preserve of the privileged but that technology and technologies of control and governance do not simply emanate from the core and proliferate to periphery but that technology is a way of approaching or approaching the world and how objects and people are used in the world. It is important for us to be open to technologies in that broad sense; not just coming from the core but also from the periphery. Technologies might pop up from anywhere.

Can: *Recent works that appear in Security Dialogue, of which you are the editor of, as well as a lot of the recent security-oriented publications, focus on effects, or externalities of "security technologies." Do you think this is limiting our understanding of security? To give an example, your anecdote about Kenya and the use of cellphones, of course this issue has a security component. in that Kenyan authorities must be investing in on developing their means to monitor cellphone usage as part of their intelligence gathering efforts, but there is also an economic logic behind these actions of everyday Kenyans, which I find that a lot of security focus either ignores, or cannot capture, or avoids all together. Do you think that this lack of communication between International Political Economy and Security Studies is undermining our potential for developing a more complete understanding of technology?*

Mark: I think there is a similarity in Foucault, Bourdieu, and Latour's statements about what drives their research agenda. As I mentioned earlier, my self-understanding is about curiosity-driven research. Foucault has a great line, that always stuck with me where he says that all of his research projects started with a personal experience that he then pursues and follows. When he was a student in between the doctors and the patients that gave him a really unique view on the operation of the clinic. That really resonated with my own sense of being an immigrant and being of two culture, but between two cultures. I read that same kind curiosity, and the kind of suspension of disbelief, or suspension of same kinds of pre-existing modes of analysis in the work by Latour and Law, where in each of the cases, they say "let's focus on this technology, or on this object and see what happens." Rather than making an assumption that a case is about security, or about economics, or surveillance, or sovereignty that we pick up and follow a string no matter where it goes and as we see those influences and record and take them. We need to be reminded of this because of the disciplinary structure of our work life, but intellectually that's the most honest and most productive perspective; start with the curiosity and intuition and that follow it wherever it goes, brining in security, democracy, surveillance, rights rather than sort of rendering all of those other considerations secondary in order to prove one's point. What strikes me about all of those intellectuals is that they all make the same kind of argument: they start with curiosity, they start with intuition and build up a compelling story.

Can: Thank you very much.

Mark: Thank you so much.

WORKS REFERENCED

Barber, B. (1992). Jihad vs. McWorld. *The Atlantic Monthly, 269*(3), 53–65.

Bennett, J. (2009). *Vibrant matter: A political ecology of things.* Durham and London: Duke University Press.

Connolly, W. E. (2013). *The fragility of things: Self-organizing processes, neoliberal fantasies, and democratic activism.* Durham and London: Duke University Press.

Salter, M. (Ed.). (2015). *Making things international, Volume I: Circuits and motion.* Minneapolis: University of Minnesota Press.

Salter, M. (Ed.). (2016). *Making things international, Volume II: Catalysts and reactions.* Minneapolis: University of Minnesota Press.

Salter, M. B., & Mutlu, C. E. (Eds.). (2013). *Research methods in critical security studies: An introduction.* London and New York: Routledge.

CHAPTER 6

The Meta-Power of Technology

Sarah Logan in conversation with J. P. Singh

Abstract Professor Singh's first book, *Leapfrogging Development?* was animated by the puzzle posed by what appeared to be a sudden focus by development practitioners on the telecommunications sector. In this discussion, he outlines the ways that initial puzzle led to a career driven by seeking to understand the transformational impact of technology at the global level and on the discipline of International Relations. In doing so, Professor Singh offers a striking history of the emergence of the academic study of technology in International Relations, from workshops to conferences, publications and personalities.

Keywords Development · Political economy · International relations · Power · Technology

S. Logan (✉)
University of New South Wales, Sydney, NSW, Australia

J. P. Singh
University of Edinburgh, Edinburgh, UK

55

Sarah: I wanted to start by asking, on reflection, how do you feel that technology has been present in your work? And what kind of an influence has it had on your thinking about IR?

JP: Well, for me technology has been present in my work ever since I started working on my dissertation. So that's been one of the major themes of my research. When I was looking for a dissertation topic, that was late 1980s, the developing world was just beginning to prioritise technologies like telecommunication. So my puzzle really was why telecommunications in a world in which we were used to thinking of development and poverty issues in terms of basic needs—water, food. Now all of a sudden you had developing world starting to prioritise telecommunications. It seems so basic now, but it was puzzling then. Why would countries which had—at that time, for example, India had almost 50% of the population under the poverty line—why would they prioritise telecommunication? That led me to thinking about telecommunications for sure—why would you prioritise a high-tech issue—but I also began to think about why would countries do it differently.

So my dissertation, which resulted in my first book, *Leapfrogging Development?*, was really a kind of a comparative political economy of telecommunications restructurings around the developing world. But then that brought up a whole host of questions, because it was broadly situated in international political economy. The overarching question was, 'Why aren't we, in international relations, paying attention to technology? I invited James Rosenau, who had just written *Turbulence in World Politics*, to join me in convening a workshop at that time, which brought together some of the cutting-edge scholars at that time in the mid-90s who were beginning to pay attention to technologies. I was at the University of Mississippi at that time, and we met in Memphis, Tennessee. That resulted in a co-edited volume with Jim Rosenau, called *Information Technologies and Global Politics*.

The two projects helped me start to think about technology in a comparative as well as an international fashion. And also, more importantly, to look at how IR had treated technology in the past in its analyses—so the usual "isms". For example, How did realism deal with it? So Gilpin's *War and Change* has a conception of technology, mostly about expanding empires and capabilities, and, liberal internationalist analysis as you know has a conception of technology. I wouldn't say that IR had ignored technology, but it was just sort of assigned the same kind of value that we had for many other kinds of transformational changes. At those Memphis workshops is where I began to think

that IR was really not paying attention to the transformational potential of technology as some other fields had. So sociologists were talking about the information society, and economists had been speaking of
knowledge workers and the knowledge society, and yet we were sort of
behind in IR to think of any kind of a transformational understanding.

And from there came what I started calling my notion of "meta-
power"—that technology changes the identity of the issues that we're
looking at—that security in the information age is not the same thing as
(national) security. Technology changes the epistemes by which we view
security, and also changes the identity of the actors themselves, in many
ways. Since then I've kind of stayed in that realm seeking to understand
the transformational impact of technology at the global level. I use the
word 'global' carefully, because global is local, and local is global—but
I've been always interested in things at the level of global governance
versus things at the level of society, asking 'how do we understand the
impact of technologies there'? I'm working on a project right now,
which is called Development 2.0, and it's—to me—the follow-up of my
Leapfrogging book.

Sarah: Could you tell me more about that?

JP: I wanted to look at what forms of participation come about as a
result of technologies, and ask–if we can count various forms of
participation—which ones might be more effective for development
versus otherwise. So the *Leapfrogging* project looked at very state-led
kind of telecom restructuring, and I really wanted examine how we
as individuals participate? What are our societal conceptions of technology? A decade ago I finished *Negotiating the Global Information
Economy*, which was really a look at how we frame global rules for what
we're calling the global information economy. So I think I have for the
most part stayed with transformational impacts and looking at those
impacts at the global and national governance levels, and now I am
hoping to look more at the local level.

Sarah: In your opinion, given your focus on global level analysis, how have
approaches to the relationship between technology and concepts like
power, security and global order developed in IR? Since the book that
you published with Rosenau in 2002, what have you seen that has actually changed in IR in this respect, if anything? And what do you think
has driven that change?

JP: One of the reasons I went to Jim Rosenau was because he was one of
these scholars who didn't mind looking at big changes. I think because
of our methodologies, because of our limitations, we gravitate towards
incremental change in IR, and he was one of those scholars who would
talk about transformational changes. And I thought that for a discipline

that really just kind of held technology constant, we needed to look at scholarship like his.

And if you look at Ernst Haas, Susan Strange, you know, Robert Cox, these are all people who really made us rethink the nature of world order, global orders, and they've asked these big questions. Not all of us, including myself, could answer those big questions. But, having asked them, they allow for the rest of us to come in and ask other questions. So in that 2002 project it was very important to get in scholars who were asking some of these big questions. Ronald Deibert was another one who contributed to the volume and he'd just finished his book, *Parchment, Printing, and Hypermedia*, and he was looking at how social epistemes changed through various ages. He anlayzed the feudal age, the medieval age and, you know, modern and post-modern age, etcetera.

There was already an impetus within scholarship by the time people like me arrived in the mid-90s. The idea was to push that further. In his article, I think in 1990—the one in ISQ—James Der Derian was talking about surveillance and things when a lot of people weren't even considering it, and—he was talking about the transformation of diplomacy as a result. These scholars paved the way for a whole host of scholars to take technology seriously. I got a position at the University of Mississippi, which was funded by BellSouth, and the department chair was offering me the position even without really interviewing me because there were so few people doing technology that they said—"BellSouth, we found somebody who can do technology and IR."

That would not happen now, you know. And, in fact, I remember the chair of the department telling me a few years later, "You know, if you'd applied four years later, there would have been a lot of competition." I said, "You know, I didn't apply. You asked me."

Now we also have a section of the ISA called Science, Technology, Art in International Relations. I played a leadership role in the Science, Technology, and Environmental Politics section of the APSA and technology scholars there we would almost meet as a support group. We would say to each other, "Oh, nobody takes us seriously. We have to publish in interdisciplinary journals, and we don't get published." And some of that is still there. But back then it really was an urgency. There really were no journals to publish in. One thing I helped to start was—well, it was already there but I just rebranded it–was the *Review of Policy Research*, which is now the journal for the politics and policy of science and technology. The policy studies organisation came to me and they said, "Can you help us with a journal?" I said, "Only if you let me do science and technology."

Overall, I just think that it's become all right now for scholars to come in write their dissertations and write their books on technology: there are so many of them now than in the past. I really want to pay homage to these big thinkers who came before people like me, who would ask these questions. With Jim too, I just said, "Jim, you keep talking about technology and turbulence in global politics, but you really don't say much about it." So he said, "Well, what should we do?" And that's how the idea for the conference came about. I said, "Well, let's do an APSA workshop," and then after that I got some funding to do the book and we convened another workshop in Memphis, Tennessee.

Sarah: The pace of change of research on these issues has accelerated very recently, given that you have been working—and others have been working on these issues for quite a long time. What do you think has initiated that acceleration, if, indeed, you agree it exists?

JP: it's a good question. There are factors internal and external to the discipline. Let me start with the external ones. The external ones are the same kinds of factors which I think allowed Wendt to write his book or Jim Rosenau to write his book, or Cox to write his book on world orders, which is that by the late 80s and the early 1990s there was a change in global politics. At that time Berlin Wall had come down, the Soviet Union had broken apart, and we had the Clinton era, and Gore talking about information technology. There were new things on the horizon. So I think those factors pushed people to puzzling a little bit more about global politics in general.

Sarah: I wanted to move to some of the major conceptual shifts that we've seen in the discipline. And given your work on meta-power, I'd love to hear what you have to say about how approaches to the relationship between technology and, sort of, these big, order-driven concepts like power, security and global order has actually developed in IR.

JP: Well, for me the conception and operationalization of meta-power has been a demonstration of constructivism. Whether we like it or not, we're all constructivists now at some level, even when Krasner writes his book on sovereignty as hypocrisy. It's the biggest homage ever to constructivism, in saying: sovereignty has always been constructed and great powers construct it. Well, sure, but isn't that also saying you're a constructivist now.

So I think these big questions allowed people to think, why did we take things are normalised? What's behind all this? I think Cox was asking similar questions—that the materiality of factors has some kind of an origin. And, similarly, in the disciplines from which IR has borrowed so much—those that deal with Foucauldian analysis or what at one time pejoratively got called that more reflective tradition—but I think very critical traditions are structuralist traditions, discourse-based traditions, I think those have come into IR in parallel with looking at technologies. And other questions have continued to come up which have made us look at technology. For example, one of the people I didn't invite to my book with Jim Rosenau but is one of my dearest colleagues now is Susan Sell. She was writing about intellectual property in the mid-90s, but at that time I just didn't know about her work when I convened the conference with Jim Rosenau. But those are technological questions, and it just sounded so odd that somebody would write on intellectual property. None of us had heard of it! And now look at the number of people working on intellectual property, on privacy and surveillance! I invited Karen Litfin to come to that initial workshop—because she'd been working on satellites and surveillance, and she contributed to the volume with Jim Rosenau.

Within the bigger question of technology there are so many different strands. But the big, big question that you asked me would have to be that technology has really made us rethink why we thought the world the way it was, to the extent that even a realist like Krasner has had to resituate new realism in that paradigm and say, "Sovereignty has always been hypocrisy." In other words, it's always been somebody's idea. It's not just that technology led to that, but the people who had begun to ask those questions had paid attention to technology when asking them.

Sarah: So the hurdle had been jumped, if you like …

JP: Yes.

Sarah: Do you see that the influence of technology on the underlying concepts of IR as clustering around any particular concept? So like power, or order? Or do you see it as a—kind of a broad-based infiltration, if you like?

JP: That's a very good question. I'm going to give you a brief disagreement that I had with Jim when we were doing this book on *Information Technologies and Global Politics*. If you look at the subtitle—I think it's "*Power and Governance*". He said, "No, not power."

Sarah: Really?

JP: This is at the workshop. And he just felt like political scientists were obsessed in their obsession about power, and we laughed at him. He said that because of that obsession with power they didn't ask some of the more substantive questions that needed to get asked. Quite obviously I not only disagreed with him, but that book emerging from the workshop was the book in which I coined the concept of "meta-power". But with him what was more important was not what is power, but what power does, where power comes from. So he just felt that power, by itself, was a useless category, because it had this endogeneity problem. You couldn't say what it does unless you looked at the outcome. And so he was pushing us more to think about the effects of power. And I appreciated that, because what he was saying was that our obsession with power confounds the causes and the effects, and that if we're going to speak to power, then we should try to keep those separate.

In general, I would say there are three things going on. One is we look at technology and we justify it within the limits of whatever ideological paradigm we belong to. So realists are going to come back at power. It's always going to increase state capabilities. Drezner does that. He does it very nicely in *All Politics is Global*, in that he says we do have multilateralism, etcetera, but at the end it enhances the power of one chief actor, which is great powers.

And, you know, liberal internationalists are going to say, "This leads to forms of cooperation." I'd like to think that in my brand of liberal internationalism I'm very nuanced. That's what I did in *Negotiating the Global Information Economy*. But somebody looking at my scholarship could say, "Well, at the end of the day you've delivered on what liberal internationalism has always argued—that we can cooperate." Drawing on critical scholars, technologies are always enhancing the power of neoliberal actors these days. No matter what technology comes about, somewhere it enhances the power of neoliberalism. And then within that paradigm are the resistance actors—those who are using those technologies to resist neoliberalism. So, at some level, an unfair critique of all of these paradigms—which include me—would be that we keep reproducing the same thing. That's one.

And the second one would that we as a discipline tend to fail to focus on those who pay attention to specific issues. I really admire them. For me, somebody like Susan Sell, who not just knows intellectual property issues inside out, and also knows the law about it. She knows the technologies of intellectual property and where they're getting applied. And these are people who I think at some level just have a deep-seated affection for empirical stuff. And then they emerge from them and they try to make sense of things. A few empirical models may also still be critiques of neoliberalism. For example, Susan doesn't use that term, but she's very uncomfortable with the capitalist compact with intellectual property. In her Cambridge book, for example, she came up with this whole notion of structuring agency, because she was struggling with these questions of pro-IP versus anti-IP—and she's an Ernie Haas student who looked for deeper interpretative meanings in his work.

Then I think there is a third set of scholars—and that's been very important for us in STAIR. They are scholars who are in IR. IR itself is quite canonical in those isms and these scholars were educated with those isms but they also came from some other disciplinary, epistemic understandings. And that has just brought in issues that are so important—gender studies and technology, for example. I don't know how many realists have ever read Donna Haraway, but the fact is that we owe it to FTGS to have brought those questions to IR.

In that vein, of bringing scholars together, what I can claim for STAIR is how many IR scholars would have ever thought of the A for Art. When we were getting the STAIR charter approved, especially the North American scholars thought, "Why A with ST?" But to us it was human design. I remember the governing council meeting the non-North American scholars, that was not a huge issue for them. But I just think they think of it differently: those who've read Bruno Latour in Europe don't have a problem going from ST to A. Whereas those who've already read—I won't name favourite realists, liberals, etcetera—have a problem there. So those would be the three ways that IR has dealt with these big, conceptual shifts over time.

But If I had to really come up with an answer, in terms of major shifts on concepts like agency and power, I would say no: it hasn't happened yet. But its not that IR has just dug in its heels–that's not valid, because we have changed since the 1980s to today. Even in Drezner's book on *Politics is Global*, he deals with internet governance. He deals with intellectual property issues. Or his book on zombies, you know, he has the chapter on gender. And so things do change.

The way I saw it in that volume with Rosenau was mainly about asking, 'In looking at technology, have we revised our major conceptions of power, which were mostly instrumental conceptions' I think Marxist, liberal, realist conceptions are mostly instrumental. Either somebody's power to do something is constrained or it's enhanced. That's what technology does? Not really. But at that time, if 90% of the scholars were saying that, I would say now 60% are.

Sarah: That's a wonderful answer. Thank you. So my next question is, to what extent can our ideas about the role and implications of industrial-age technology help us understand the challenges and opportunities of the information age? So in what way can our understandings of previous technologies help us? Or can they not?

JP: I think in a way both questions are the same, is the way I've dealt with it. We had these transformations in the past. What were they? So when the railroad came—or the railway came, what did it do? Well, let me take the railways as an example. Albert Fishlow, the economist, wrote

Railways and the Antebellum Economy. He argued that railroads didn't carve out the West because it was not the frontier. Because there'd been this thinking that railroads carved out the west and people followed. So it was a supply-push argument. He reversed it and said, "No. Railroads followed demand. People had already gone out there. They were logging. They were mining. They were doing all these things, then the railroad followed."

So by looking at a technology like the railroad, he was able to take a sort of a dominant conception of technology that is transformational in that it alters the supply equations and the railroads follow. That was within the dominant paradigms of economic history and neoclassical economics. Then you take somebody like James Carey working on the telegraph, and he ends up speaking about technology. He says that when the railroads came, they needed standardised time. At that time in the US all the cities fixed their own times. And so there was this huge battle, because the city square had the clock on top, and that was the time for that city.

And then who backed the square? All the authorities in the city, including the church. So the railroad, for them, was the devil's engine. Because now the mechanical devil was coming and it was going to change the way we looked at divine time. In the end, we got standardised time zones because the railroads wanted standardised time zones, and because you couldn't go from one place to another if you couldn't send signals that the train was coming. Therefore, in a way, the notion of time changed. If I take those two together, Fishlow versus Carey, one was able to take what existed within theory—existing theories— and say, "Don't be so supply push. There's a demand-led argument." Then the other one was really pushing at the frontiers of, "This changes our thinking about how we see something as basic as time," for example. Another example Carey gives is how arbitrage used to be between two cities, because you didn't know what the product was selling at in another city. When the telegraph came, you immediately knew what the price of corn was in Chicago and in St Louis. So we went from spatial arbitrage to temporal arbitrage. So futures trading started when the telegraph came.

Returning to those historical questions allows us to say, how—what are the kinds of questions we asked about the industrial age? How were they answered? How did they answer it in terms of the discipline of economics? And when you loosen those conditions, how do they help us ask more transformational questions? I think for the information age or the post-industrial age, that's the way we should be asking those questions.

Sarah: Given your long history of scholarly involvement in these issues, what do you see as the most significant challenges and opportunities for International Relations in the context of technological change? I also wanted to move on from that to a discussion which places your concept of meta-power in the context of the recent American election results, if you're happy to do that.

JP: Sure. In fact, I think the two are related. In that meta-power piece for ISR, I distinguished it as being emancipatory at some level. And the co-editor for that volume was ISA President Beth Simmons, because it was her issue for ISR—it was the presidential issue. What she kept pushing me on was the non-emancipatory potential of these technologies. She's a liberal internationalist, but she said, "Why are we getting these homophily effects? Why are we getting these stovepiping effects? Why are human beings, in millions, drawn to a perspective that we would have considered marginal? And it's not." And at that time it seemed like, "Oh, maybe 10 per cent of the people are," and then we elected Donald Trump, or we have Brexit, and we have elections in Italy, Germany, France and Austria where populism is a very real issue. Somewhere along the way there's something about information technologies which are making groups cohere in particular ways and populism is on the rise along with xenophobia and racism. We now know that Facebook not only played a role in propagating fake news, but that bots and algorithms helped create support for right-wing populism using Facebook data. Furthere popoulism has an international dimension through social media. Facebook increased the potential of these sort of transnational communities to find each other. So it wasn't just Nigel Farage helping Donald Trump. It's like Donald Trump supporters in the Midwest might be able to empathise with the Farage supporters in the English Midlands.

We're going to have to puzzle about what's going on with these technologies. We will need to venture into unknown spaces. Do you know Eszter Hargittai? She's a sociologist at Northwestern. She argued in 2016 that social media didn't play as much of a role in the elections as we like to believe for particular demographics.

She's basically said that the people who voted for Trump and the people who voted for Brexit—she hasn't said Brexit, but she's talking about Trump—are the ones who are less likely to be connected in social media, such as the older generations. And the ones who are most connected, the younger ones, voted for Hillary. And it's an important point in some ways. So I appreciate her insight here, because she's just saying, "Before we all jump on this bandwagon of social media caused this to happen, be careful. Because when you start looking at the demographics, the most connected populations are voting for Hillary." So I just think that those are helping to set up these parameters of this debate. That means we still need answers how particular demographics, or those connected and disconnected, are influenced in their political and cultural beliefs through social media. We have a few answers to how powerful algorithms and big data influence elections but we have a long ways to go to understand the unexpected polarization our politics induced from the very technologies that we thought were emancipatory and democratizing.

Right now technologies are posing a huge question for us, in terms of democracy, solidarity, transnational movements and how ideas are conveyed. And if we're looking at the 2016 elections and votes, which includes Brexit, I think those are the kinds of puzzles we'll have to attend to in the future.

Because technologies allow for the non-progressive forces of the world to cohere, you don't have to be particularly enlightened to use these technologies. And since Plato we've thought of technology as enlightening. Through the parable of the caveman, technology had this enlightenment effect. I would, by the way, also say that there are theorists of technologies who warned us about this. Was it Leon Kass? He's a political theorist, and in one of his essays he speaks to how the condition of democracy is that the citizenry must be conscious and enlightened.

And we've moved away from the old conceptions of democracy— in the very basic sense of saying, well, only the landlords are enlightened, so people who don't own property, we won't let them vote: we've broadened that sphere, including suffrage or allowing minorities to vote and so on. But Kass says that now we have another crisis on the horizon. In his book he speaks to how more than 50% of the American population is on some kind of mood-altering drug or the other. He says, what does that mean for the condition of democracy? He's a conservative political philosopher, but I think he's asking a fundamental question. When we've been altered—Donna Haraway is asking the same question in Cyborg—when we are altered by technology, what does it mean for the political systems we have? What does it mean for global governance? What does it mean for everyday actions or society? How do we cohere with each other?

Sarah: On that note, one of the most interesting outcomes of the election in this context is the role of algorithmic power and bots, and non-human practices of technology, if you like, and how we might understand their role. Is that a type of power? Is it a type of agency? Do you have any thoughts on the role of non-human agency in this?

JP: There is a chord within those technologies. This is where we can learn from STS scholars that technologies are not neutral, which again brings me back to Jim Rosenau. Another disagreement that I had with him was, he says in his essay, technologies are neutral. And I remember going—taking him for lunch at one time in Washington DC and saying, "Jim, I want to convince you that how technologies are not neutral. They have social relations built into them." And he—so he said, "You need to think about first-order effects and second-order effects and third-order effects, in terms of these – what technologies embody." I never quite could persuade him. But I think he persuaded me a little more than I persuaded him. For those of us who grew up with STS, at some level, we don't take technology to be neutral. We don't take algorithms to be neutral, even when they are not consciously created for a particular outcome which ends up taking place—the consequences of data driven advertising revenue and so on.

Now, a critical studies scholar would come at it and say, "I told you so. You know technology is not neutral." But I would say is that Jim would be right in that he would say, "Well, only money by itself is not a nefarious enterprise." So that's not a first-order effect. Then he would say, "A second-order effect is that this has led to people only doing these things." And a third-order effect might be that it builds. There may be also ethical incantations of this, that it allows for certain conceptions of evil to come about in society that didn't exist before.

Consider what Cambridge Analytica did in perhaps acquiring data illegally through Facebook and then building profiles of people's political preferences so they could be targeted with particular messages. There's nothing inherently wrong with using big data to reach particular audiences. Jim Rosenau would argue that technology is, therefore, neutral and can be used for positive and negative purposes. But the entire scale of the operation here, the possible illegal breaches of privacy, and that the targeting led to extreme social divisiveness and populism leads us all to take pause and reflect beyond technological neutrality and question how it shapes human consciousness and beliefs.

Sarah: That's fascinating—the idea of evil being altered by changes in the conception of agency.

JP: Yes. Actually, people like Hannah Arendt and Heidegger, were asking these questions about technology and evil in the 1930s.

Sarah: That seems like a very clear point on which to end our discussion, if you're happy to finish there. Thank you so much for your time!

JP: Thank you! But we end on a sad note.

NOTES

1. Feminist Theory and Gender Studies (FTGS) is a section of the International Studies Association.
2. Simmons, B., & Singh, J. P. (Eds). Special Issue: International Relationships in the Information Age. *International Studies Review*, Vol. 15, No. 1 (2013).

WORKS REFERENCED

Carey, J. (1989). *Communication as culture: Essays on media and culture.* New York: Unwin Hyman.

Cox, R. W. (1987). *Production, power, and world order: Social forces in the making of history* (Vol. 1). New York: Columbia University Press.

Deibert, R. (2000). *Parchment, printing, and hypermedia: Communication and world order transformation.* New York: Columbia University Press.

Der Derian, J. (2009). The (s)pace of international relations: Simulation, surveillance, and speed. *International Studies Quarterly, 34*(3), 295–310.

Drezner, D. W. (2008). *All politics is global: Explaining international regulatory regimes.* Princeton: Princeton University Press.

Fishlow, A. (1965). *American railroads and the transformation of the ante-bellum economy.* Cambridge, MA: Harvard University Press.

Gilpin, R. (1981). *War and change in world politics.* Cambridge: Cambridge University Press.

Haraway, D. J. (1985). *A manifesto for cyborgs: Science, technology, and socialist feminism in the 1980s.* San Francisco, CA: Center for Social Research and Education.

Rosenau, J. N. (1990). *Turbulence in world politics: A theory of change and continuity.* Princeton: Princeton University Press.

Rosenau, J. N., & Singh, J. P. (1990). *Information technologies and global politics: The changing scope of power and governance.* Albany: State University of New York Press.

Sell, S. K. (2003). *Private power, public law: The globalization of intellectual property rights.* Cambridge, UK: Cambridge University Press.

Singh, J. P. (1999). *Leapfrogging development?: The political economy of telecommunications restructuring.* Albany: State University of New York Press.

Singh, J. P. (2008). *Negotiation and the global information economy.* Cambridge, UK: Cambridge University Press.

Singh, J. P. (2013). Information technologies, meta-power, and transformations in global politics. *International Studies Review, 15*(1), 5–29.

Wendt, A. (1999). *Social theory of international politics.* Cambridge, UK: Cambridge University Press.

CHAPTER 7

Culture, Diversity and Technology

Constance Duncombe in conversation with Christian Reus-Smit

Abstract Christian Reus-Smit is one of the leading experts on international theory, history, and international law. He has pioneered new conceptualizations of individual rights and political legitimacy in the development of international orders, generating innovative debates around concepts of legitimacy, power and social and political theory at the intersection of international relations. This chapter details a 30-minute conversation between Reus-Smit and Constance Duncombe in his office in the School of Political Science and International Studies at The University of Queensland in August 2017, which explored how technology has both informed his work and how he understands its social and cultural underpinnings. In this conversation Reus-Smit discusses technology as a material artefact and its role in transformative change of the international order. Key here is the insight that how we conceptualize

C. Duncombe (✉)
Monash University, Clayton, VIC, Australia

C. Reus-Smit
The University of Queensland, St Lucia, QLD, Australia

C. Kaltofen et al. (eds.), *Technologies of International Relations*,
https://doi.org/10.1007/978-3-319-97418-7_7

technology, including social media, matters for how we understand its material and ideational power. Reus-Smit's principal innovation is to see technology not just as a social artefact, but also as congealed ideas.

Keywords International order · Power · Constructivism · Transnational politics · Norms · Culture

One of the leading experts on international theory, history, and international law is Christian Reus-Smit. He has pioneered new conceptualisations of individual rights and political legitimacy in the development of international orders. His extensive corpus of published work has generated innovative debates around concepts of legitimacy, power and social and political theory at the intersection of international relations (Price & Reus-Smit, 1998; Reus-Smit, 1999, 2004, 2007, 2012), including examinations of nuclear strategy and arms control (Reus-Smit, 1989), international institutions (Reus-Smit, 1997), human rights (Reus-Smit, 2001, 2011, 2013), and culture and international orders (Reus-Smit, 2017). In our 30-minute conversation, which took place in his office in early August 2017, we discussed not only whether considerations of technology have explicitly or implicitly informed his work, but also how technology as a material artefact is socially and culturally proscribed. Technology matters to international politics. Technological change has been key to transformations of the international order, in terms of how we conceptualise ourselves as political units as well as shifts in material processes and relational power dynamics. Yet technology, and its role in fomenting wide-ranging political change, cannot be fully understood unless we examine the social and cultural relations within which it is embedded. As this conversation with Christian Reus-Smit illustrates, how we think about technology gives us much greater insight into its power, both in material and ideational terms. Before we can measure the impact of technology on as a discipline, or world politics in general, the reasons behind the emergence of technology—including social media—must be acknowledged.

> *Constance: Thank you very much Chris for the interview, I'm very excited about this. To start off more briefly, what does the idea of technology mean to you?*
> *Chris: Technology for me is materially embedded ideas, and as a consequence I don't see technology simply as material stuff that you do things with. It's a manifestation of particular ideas, often within*

a material medium. What this means is that technology is always socially embedded, it is embedded in the ideas of the innovators that create technology, and those ideas are always embedded within wider social frames. I'm someone who sees technology very much as existing within an ideational context, as being the place in which technology emerges and in which it matters.

Constance: How do you feel technology has been present in your work, and what kind of an influence has it had on your thinking about IR?

Chris: The most honest answer to that question is that it has been present in my work at a very subconscious level. Unlike someone like (Phillips 2016; Phillips & Sharman, 2015), for example, who in his work technology and changes in technology are one of the variables in his story, you won't find that in any of my work. But when I think about it, technology features in my work as part of background social change. For example, if you look at the work that I did on the rise of the modern international order in the 19th century, I talked about how ideas about legitimate statehood, and about political legitimacy more generally, emerge in a context of multiple social transformations. You have transformations wrought by the industrial revolution; you have social transformations at the level of ideas about the relationship between economy and society; and you have social transformation in a shift in understanding of society from an understanding that is essentially holist, which begins with society and then understands the individual within that context, to one in which the individual is the primary building block of society. As soon as you talk about the industrial revolution, or ideas about economy and society, technology clearly played a very important part as an enabler of the kinds of economic and political transformations that we're talking about. But at the same time, those technologies were embedded ideas. The technology of the production line that was so important in the industrial revolution was only possible when human social processes could be imagined as comprising into individual units, which could then be reconstructed in economically efficient social processes. I'm a radical sceptic about any argument about technological determinism. I think that technologies always emerge in the context of ideas: indeed, they are embedded, congealed ideas.

Constance: Your earlier work looks at arms control and nuclear strategy, whereas now you're focussing on questions of international orders, diversity and transnational actors. Does this change in focus mean that you've moved away from technology? Or has your conceptualisation of what technology is changed?

Chris: First of all, very good for discovering that I did early work on nuclear strategy and arms control, not a lot of people know that and I'm not sure that I should admit it! The honest answer to your question is that even when I did work on arms control, nuclear technology, and nuclear strategy, the question of the technology itself was not especially important to me. I was writing towards the end of the Second Cold War, when the intensification of the nuclear arms race was incredibly important. But what particularly interested me was the way in which strategists thought about the utility of nuclear weapons. Many of the developments that were occurring in nuclear technology at that stage were reflecting a tension in ideas about the utility of nuclear weapons: between the conventional mutually assured destruction position and a war-fighting position (the idea that nuclear weapons might be usable under certain battlefield conditions). That was the central tension in the development of nuclear weapons at the time, and it was these arguments about the utility of nuclear weapons that were in part driving the technology.

With the work that I'm doing on cultural diversity, I'd say two things. One is that culture has a material face. It's not something that exists just at the level of ideas, if there's such a thing, but it's also embedded and congealed in material artefacts. Again, because I see technology as congealed ideas, the actual boundary between the material and the ideational in cultural forms is not clear-cut. The work I'm doing on cultural diversity can connect to questions of technology through that idea of material artefacts. At the same time, though, I'm trying to think about culture as inherently diverse, which means that cultural ideas and practices are manifest in highly contradictory ways, in discourses, in practices, in material artefacts. As much as anything else, what interests me is the manifestation and expression of that inherent diversity. I tend to focus much more on the discursive expression of culture, or its expression in practices, partly because it's much harder to study the material artefact side of that process at the international level. It would be fair to say that, again, like the nuclear work, the central concern of my work is not the technology itself, but it's part of the broader context.

Constance: How does IR understand the relationship between technology and concepts like power, security and global order?

Chris: For realists, technology is usually listed among the material sources of state power. Yet technology itself reveals how problematic narrowly material conceptions of power are. If what I said earlier is true—that technology is essentially congealed ideas—then it is in reality deeply social. Recognizing this resonates with recent shifts in IR, where the social sources of power are receiving far greater attention. Diverse schools of thought are now probing power more deeply, and pointing to its very strong social dimensions and social ingredients. And when we think about technology, technology is something that is as social as it is material, and therefore we need to think about it in a more sophisticated way in its relationship to that broader conception of power.

Constance: What are the challenges and opportunities that technology provides in IR, both as a discipline and also in practice?

Chris: There is still an impulse in the field that whenever you talk about technology it's done in a fairly deterministic way. There is a sense that technology emerges through some process, but IR scholars have treated these processes much as they treat the sources of interests. We know that interests come from somewhere, but it is assumed that we can bracket where they come from and just treat them as given. Technology is a little bit like that, in that we bracket the social formation of technology, take particular technologies as given, and then look at their effects. It is crucial that we avoid this tendency if we really want to understand the relationship between technology and politics. We have to take the notion that technology is congealed ideas seriously, and this will direct us to the broader ideational context in which technology emerges. Ideas aren't free floating, so that involves looking at the social processes by which particular ideas get mobilised and transformed into technologies.

The issue for politics is that technologies don't just rise, or fall or survive or die, take off, whatever, as a consequence of the genius of those people that invented them, or the kind of dynamics of capitalism. They also depend very much on the kind of political regulatory environment in which they emerge. That environment shapes the field of possibility. Imagine technology in the middle between a set of social forces and a set of political or institutional structures that provide regulatory frameworks that may or may not be directed at that particular technology. The fortunes of that technology depend on that place, between social forces and political regulatory environments. Sometimes social forces drive technology ahead of regulatory environments, sometimes regulatory environments are ahead of the technology. Sometimes regulatory environments run ahead of technologies, sometimes regulatory environments that have nothing to do with the technology concerned shape the emergence and development of a technology. It's that space we need to be thinking about when we consider the impact that technology will have on politics. I'm deliberately not talking about 'international politics' or 'domestic politics', because I think we're at a stage now where the boundaries between those things are really quite blurred, for the purposes of understanding the impact of technology.

Constance: You've recently joined Twitter. How do you feel that it has changed—or not—the way you engage with the everyday, and IR? Do you notice a change in the way you perceive the world, or people's work?

Chris: I admit to being a cynic about Twitter. I did join Twitter, but I use it sparingly, both as a source of information and also to get the word out about things. If you want to reach a particular audience, it's a good vehicle. But my overall impression is that Twitter is a medium that's produces an enormous amount of noise. One of the things that worried me about it (when I joined) was that I would be drawn into it, and that I would spend my time looking at Twitter. I was disabused of that after about three days, where I discovered that ninety percent or more of what's on Twitter is noise. And there are interesting social and psychological questions to be asked about why otherwise brilliant individuals feel the need to air every stray thought they have. More seriously, I think Twitter is a very useful medium for gathering and circulating information. But there's also a curious social function that it is performing between people; in how they're imagining their communicative communities and also what communication actually means in that space.

Constance: Do you think technology such as social media is beneficial or problematic for conceptualising world politics? And what might the use of such technology by transnational actors mean for current relations of power?

Chris: If we go back to an idea that was there in Robert Gilpin's work (1981, 1987), and was later elaborated by (Buzan & Lawson, 2015) and others, we see that a key variable in any international system or order is its relative interaction capacity. It doesn't take much to see that social media technologies are transforming the interaction capacity within the modern international order, and that this is greatly expanding the field of politics globally. Yet the technology itself doesn't determine the content. And so the fact that Donald Trump uses this technology to reach a particular audience tells us something about interaction capacity, tells us something about the changing field of politics in which he's operating, but there are deeper questions about why there is a community responsive to these ideas in the first place, which the technology doesn't give us the answers to. We really have to avoid the idea of assuming very simple causal arguments, because there's always a wider social field in which that technology is operating and being used; it's essential to any explanations of the rise and empowerment of actors and how they mobilise. The fact that ISIS uses Twitter and social media is really important for their ability to mobilise, but their ability to mobilise is also determined by the content of the ideas that they're mobilising, and how those ideas resonate in larger social contexts where those ideas are meaningful.

Constance: One final question: where do you think technology will take IR?

Chris: One way to approach this question is to ask how technology shaped the discipline of IR to this point. There have been clear technological developments that affected the discipline: the advent of industrial warfare being one. The nuclear stalemate between the superpowers, or the existence of nuclear weapons, clearly had an effect on how the field theorised stability and instability. Then of course there've been all the technologies that have been enablers of IR scholars, which will continue to transform over time.

Another approach would be to ask whether technology is likely to change the basic assumptions of the field. And that I think depends on how narrowly you draw those assumptions. If you're somebody who believes that IR is about relations between political billiard balls, then I think the world is going to change around that conception in a very challenging way. We've been wrestling with that for a long time, and an argument can be made that the earliest IR thinkers, before there was a coherent discipline, in fact didn't assume such a world. They were working in a world that was far more heterogeneous politically, in terms of structures and units of political authority, and they were able to reflect on what we now call international politics in that context quite well. Since the 1970s, there's been a large body of literature that has really challenged the state centrism of the field. So, I see the field as being not so wedded to that. States and sovereignty are likely to remain important, but we will need to think about these as embedded within a much more complex transnational political environment, in which many of the things to do with technology that we're talking about, like the ability for transnational mobilisation through the use of social media, will be important. Far from being a problem, this is what makes contemporary world politics exciting and interesting. My sense is that, yes, technology is going to be important, but it has to be understood within this complex global social environment. IR as a field will thrive in that space so long as it holds on to its central interest in politics, and so long as it doesn't fetishize existing theories.

Works Referenced

Buzan, B., & Lawson, G. (2015). *The global transformation: History, modernity and the making of international relations*. New York: Cambridge University Press.

Gilpin, R. (1981). *War and change in world politics*. New York: Cambridge University Press.

Gilpin, R. (1987). *The political economy of international relations.* Princeton: Princeton University Press.

Phillips, A. (2016). The global transformation, multiple early modernities, and international systems change. *International Theory, 8,* 481–491.

Phillips, A., & Sharman, J. (2015). *International order in diversity: War, trade and rule in the Indian Ocean.* Cambridge: Cambridge University Press.

Price, R., & Reus-Smit, C. (1998). Dangerous liaisons? Critical international theory and constructivism. *European Journal of International Relations, 4,* 259–294.

Reus-Smit, C. (1989). Arms control, nuclear strategy and Australian foreign policy: The Fraser years. *Global Change, Peace & Security, 1,* 57–73.

Reus-Smit, C. (1997). The constitutional structure of international society and the nature of fundamental institutions. *International Organization, 51,* 555–589.

Reus-Smit, C. (1999). *The moral purpose of the state: Culture, social identity, and institutional rationality in international relations.* Princeton: Princeton University Press.

Reus-Smit, C. (2001). Human rights and the social construction of sovereignty. *Review of International Studies, 27,* 519–538.

Reus-Smit, C. (2004). *American power and world order.* Cambridge: Polity Press.

Reus-Smit, C. (2007). International crises of legitimacy. *International Politics, 44,* 157–174.

Reus-Smit, C. (2011). Struggles for individual rights and the expansion of the international system. *International Organization, 65,* 207–242.

Reus-Smit, C. (2012). International Relations, irrelevant? Don't blame theory. *Millennium, 40,* 525–540.

Reus-Smit, C. (2013). *Individual rights and the making of the international system.* Cambridge: Cambridge University Press.

Reus-Smit, C. (2017). 'Cultural diversity and international order', International organization. In T. Dunne & C. Reus-Smit (Eds.), *The globalization of international society* (pp. 1–35). Oxford: Oxford University Press.

CHAPTER 8

Experts, Matters and Actor-Networks

Malcolm Campbell-Verduyn in
conversation with Tony Porter

Abstract In this interview, Tony Porter discusses conceptualizations of technologies and their usefulness in understanding the development of international industries, systems of accountability, as well as in overcoming public-private and constructivist-rationalist dichotomies. The interview also discusses the incorporation of Actor–Network Theory in IR and IPE, as well as the governance roles of digital technologies such as Big Data and their associated algorithms.

Keywords Accountability · Actor network theory · Big Data · Knowledge · Global finance · International industries · International political economy

M. Campbell-Verduyn (✉)
Department of Political Science,
University of Toronto, Toronto, ON, Canada

T. Porter
Department of Political Science,
McMaster University, Hamilton, ON, Canada
e-mail: tporter@mcmaster.ca

© The Author(s) 2019   77
C. Kaltofen et al. (eds.), *Technologies of International Relations*,
https://doi.org/10.1007/978-3-319-97418-7_8

Malcolm: To begin Professor Porter could you describe how technologies impact and have been present in your work as well as have influenced your thinking about international relations and the international political economy?

Tony: Technologies have been really important for my work right from the beginning. Technology, in my mind, has two meanings. One is that it's a technique that's used systematically to achieve consistent practical outcomes. The other meaning is to refer to technological artefacts like machines. Both of those are really interesting in studying international political economy and international relations. One thing about those meanings is that technology is much more than a resource that can be possessed like gold. It also has the capacity to structure human activities. You can think of the way a machine or a manual requires humans to be organized around it in a particular way (Callon, 1991). Or the way that codified knowledge can structure an organization or the way people do things. So I've really been interested in those aspects of technology from the very beginning. First in the way that they differentiated industries and firms so that some firms were extremely accomplished at generating new technologies and therefore able to dominate markets and structure them in a way that advantaged their own ability to make profit. The countries in which those firms were headquartered, if they were multinational, benefited from those technological advantages as well. And then also the idea that technologies mature. There's often technology cycles. In the early stages of an industry the leading firms can really dominate the industry. You can think of the role of cotton in the industrial revolution where some firms in Britain managed to dominate the whole world and then as the industry matures it becomes modified and that alters governance institutions and the politics of an industry. Now cotton is very mature but there's new industries that continue to innovate like with electrical machinery, which is organized quite differently. More recently I've been interested in the more general ways that technologies are embedded in documents or machines and the way that those circulate around globally and really have very important impacts on the way that global governance, global regulation or international organizations, public and private, are organized.

Malcolm: We appear to currently be in one of the cycles of technological change that you mention. What do you see being the greatest challenges and opportunities for international relations and the international political economy from these cycles?

Tony: There's so much technological change going on now. Some changes are specific to particular industries. For example, take the apparel and textile industry. In the 1960s, it was organized primarily through state-to-state relations where there were quota systems that divided up the market. Because it was so commodified and mature the technologies were widely known around the world. There was a lot of political conflict over those. As the industries evolved and they shifted towards more global supply chains with all sorts of technologies to coordinate suppliers of very large firms like Wal-Mart, which work through electronic systems to standardize and organize these supply chains. There's those types of dynamics that happen within industries. But then there's also some more generalized changes in the role of communications technologies or the mediation of communication more generally. This is really important. The Internet is an obvious example that has changed global politics and international political economy. The use of electronic technologies for controlling digital property within large organizations and their sharing of it through licenses and making money from is another example.

Malcolm: Technologies, including the examples you bring up, are often praised as solely good, especially in technological-centred media or for instance in the Wall Street Journal, Financial Times *etc. But they are also derided in more critical scholarship. In your own research how do you balance the wild optimism and wild pessimism that surround technology?*

Tony: There's a view that technological change is always good because who could be opposed to knowledge? But of course in the field of IR and IPE there's a lot of literature that's inspired for instance by Foucault, that shows the linkages between knowledge and power. There's also a large literature on the role of knowledge and competition, whether it's competition between states and other states or between firms. Knowledge is definitely a really important aspect of the inequality that develops between firms and between countries. So technology is closely linked to power because those who possess it can structure all sorts of interactions, whether they are market interactions or other types of interactions that are mediated through organizations. Of course in warfare the dominance of drone technologies and things like that are clearly connected with power. So in my view it's really impossible to generalize about whether technology is good or bad. It really depends on how it's connected to networks and mechanisms of accountability that make knowledge either public or private, that make it inclusive or exclusive, or connected to systems of power or systems of accountability.

Malcolm: How have IR and IPE approached the roles and impacts that these technologies have had? Where do you really situate yourself in the approaches that IR and IPE have taken to understand such impacts?

Tony: Technology has been present in the field of IR from the beginning. But often it's been, certainly in the earlier decades, treated as a resource that was possessed or not possessed by states. An obvious example is the possession of military technologies like tanks. There was an exception during the 1950s. There were some scholars who connected communications theory to the study of International Relations (Karl Deutsch, Harold Innis, Harold Lasswell are examples). They were much more sensitive to the autonomous or important roles that technologies can play in the structuring of international interactions. For instance Deutsch looked at the roles of communications technologies in fostering integration across borders. Innis looked at the role of particular types of products that were exported to the periphery from the core economies to see how that would play a role in constituting power relations. More recently there's been a much richer understanding of the role of technologies that draw on theorists such as Foucault, who had a very generalized interest in technologies. He talked about the technologies of the self, the technologies of power, that were not just about machines but more about the techniques for people to shape their own identities and bring them into line with systems of power. There's Actor Network Theory, which I've become very interested in, drawing on work by people like Bruno Latour, John Law, or Annemarie Mol. Actor Network Theory starts from the viewpoint that objects can play an important independent role in any type of human interaction and that in fact for action to happen it has to be carried through humans or objects like written documents or machine systems. That's been very helpful in getting clearer on the role of technologies and more social technologies like documents or electronic systems working in a way that's very complex. So it's not just a particular machine but a whole system that extends out through the way people are organized and all the things that flow into machinery and out of machinery are connected through networks. That complex type of networks and interactions I find very helpful from Actor Network Theory and I've been interested in how that's applicable to both IPE and IR more generally.

Malcolm: What are some of the challenges of incorporating Actor Network Theory into IR and IPE theory?

Tony: There's some aspects of IR and IPE theory that make it a challenge to bring Actor Network Theory into. An example is with the characterization of the field as divided between constructivism and rational choice theory, which was a prominent way of characterizing the field for quite a while. Often that was seen as constructivism being more purely idealist or about ideas and the rational choice being more purely about material things. That I think is a strange way to characterize those two approaches because there's all sorts of objects that are material but are also constructed like buildings and machines (Latour, 2003). Also there's a lot of norms that are codified in machines that then structure human activity. So I think getting beyond the tendency to create this binary separation between ideas and matter or between humans and objects is something that is a challenge with bringing Actor Network Theory into the fields of IR and IPE. But I think it's very worthwhile to do that because there's all sorts of examples where ideas and matter are integrated. A good example is derivatives markets in finance that have huge material consequences but are also very intangible: they are also about coordinated expectations and so on that are written in contracts. There are also some shortcomings of Actor Network Theory that quite rightly make some people hesitate in bringing it into the field. One is that often there have been tendencies, in the course of trying to highlight the active role that objects play in social relations, to downplay the unique characteristics of human imagination and agency. But there's no reason that this particular shortcoming of Actor Network Theory can't be set aside and its good parts be brought into the field.

Malcolm: *One of the binaries that IPE and IR have confronted is the public-private split. How does a focus on technology and also the incorporation of Actor Network Theory help in overcoming this traditional split IPE and IR have sought to really overcome?*

Tony: There's a tendency in a lot of IR and IPE theorizing for quite some time and to quite some degree in some research programmes to divide the world into quite autonomous and bounded actors and categories. That includes the public-private differentiation, so that the state was seen as a bounded and independent organization, and the market, or firms, were similarly seen as quite distinct and separate from the state and often as competing with it. Certainly I would say that a major theme in more recent scholarship is the complex entanglements between state and market. Technology is interesting because it crosses another boundary which I've mentioned already, which is between ideas and matter. If you think of a machine, it has ideas embedded in it. But it also has material character. Technologies are really interesting in looking at the way the public-private separation can be arbitrary. If you take the structuring effect of a technology like a computer programme— that can be an almost generic structuring effect, which can be present in a public organization or a private organization. This effect may exclude people, a privatizing effect, or it may be more inclusive, a more public effect. However whether this technology is embedded in a more public organization or a more private organization is really not fixed by the character of these two types of organizations. If you continue to pursue the implications that that has for ontology, or what the categories we take as being real, then it raises some interesting questions about how we define some organizations as public or private to begin with. Simply labelling all government organizations as public and all firms as private when you have a lot of firms performing public functions or a lot of governments operating as if they were private organizations benefiting a select group of people further highlights this arbitrary character of that public-private distinction.

Malcolm: This relates to issues of accountability as well as transparency. It also links to your previous mention of the complexities involved with technologies. How does the complexity of technology obfuscate the accountability that we are accustomed to in democratic society?

Tony: Technology has always been accompanied by experts. The experts that are associated with technology and claim to speak for it also can yield power and they are difficult to hold accountable in a democracy. One of the insights of Bruno Latour and Actor Network Theory is to point out the parallels between scientists who claim to speak for the physical world, in the case of physicists or biologists for example, and representatives in legislatures who claim to speak for citizens (Latour, 2004). There's representational issues that are important to bring out when it comes to experts speaking for the natural world in a way that may not be accountable. Technology also can work against accountability through its direct effects as a technology if it structures people's activities. Often the operations of a machine system are obscured to the people who are influenced by it. A good example is the algorithms that are present in a lot of programmes like Google or Facebook that may alter people's experience of the world–but how they do that is not clear (Dobusch, 2012). It becomes really important to look at how these technologies are connected to systems of accountability or not. It gets back to the question of whether technology is good or bad. You can't really say, before analyzing the system the technology is embedded in. If there are good systems of accountability embedded in the technology, it can be a good thing, but if it's run in an unaccountable way, it's not transparent, where there's no opportunity for people whose activities are being shaped by the technology to understand or shape it, then it's likely to have negative consequences from the point of view of unaccountability and democracy.

Malcolm: Which we're seeing at the present moment with the American Congress questioning the neutrality of the Facebook algorithm.[1] The key implication being that no technology is neutral- there's always normative implications involved. You've mentioned algorithms and perhaps we could draw on your own recent work on Big Data and transnational governance. One of the questions addressed in one of your previous articles is how numbers matter in global and transnational governance. Updating this question can we ask how do Big Data matter in international relations and the international political economy?

Tony: Numbers have become increasingly prominent in all aspects of global governance. This is evident in things like rankings, statistics, risk models, benchmarking, and all sorts of digital things that essentially involve numbers. All this is connected to the development and growth of machine systems, digital technologies and so on that facilitate and support these new types of technologies in global governance, the organization of private supply chains, and other transnational organizations. Big Data are a subset of that larger phenomenon, of the growth of numbers in governance. There's some distinctive characteristics of Big Data that are well recognized. One is velocity, the speed at which they are generated. So they might come from clicks on the Internet in real-time, continually adding to the data flow and data sources. There's volume, so the data could be far too large to put into a more conventional structured database. And then there's the variety. Data might come from multiple sources and be hard to integrate. So there's these very very large databases. The other aspect is that to create or use these, they are often accompanied by algorithms that draw on certain data and use them to structure, for instance, your experience when you are surfing the web, or when you are buying a product. There's all sorts of privacy issues associated with that, issues of power in terms of algorithms, and how do they shape people's experience of the world–and also power in terms of how some firms are dominating the use of these algorithms or the knowledge of how they operate. There's a great many actors without such power, whose experience and opportunities are being shaped by these, including things like job searches where firms may be using data generated by algorithms to draw conclusions about people's reliability for employment.

Malcolm: Increasingly Big Data are being drawn upon in finance to inform financial decisions and the risk models that are used in that sector. What are the implications of incorporating Big Data into the financial sphere?

Tony: Finance would seem to be one of the industries or areas where Big Data would be most welcomed because there's a huge amount of data that's generated through financial transactions and risk models, which are very central to global finance. These risk models have some of the same properties as algorithms working with Big Data since they are structured ways of thinking about things that are connected to all sorts of data. But surprisingly, in research that you, Marcel Goguen and I have done we've found that there's not as much progress as you might imagine in drawing on these technologies in finance (Campbell-Verduyn, Goguen, & Porter, 2017). Part of the issue is that it's very challenging to draw together the essential elements of Big Data systems. That has valuable lessons for power and capacity in finance, despite the common perception that finance is dominated by extremely powerful and capable actors. It is a challenge often to make systems work, which of course we saw in the 2007–2008 crisis. Those same types of technical problems that are connected with power conflicts are challenges for instituting Big Data systems. But nevertheless there are some huge investments in Big Data that will have consequences for how global finance is organized. These can either be ones that contribute to global financial stability, like the risk models that are being mandated by the regulators at the Basel Committee on Banking Supervision and other global fora. Or they can contribute to instability as with the case of high-frequency trading that has led to the flash crash of 2010, the very rapid drop in the market in the course of minutes because of the very fast algorithmic trading that got out of control.

Malcolm: That's a fantastic example and the 2013 Twitter-linked hack or 'hash' crash is another (Karppi & Crawford, 2016). *We have one last minute for a concluding comment. Was there anything you would like to add that we haven't drawn upon in our discussion?*

Tony: Technology, as I've mentioned, has been really interesting to people studying international affairs for a long time. But I think today it's connected with a broader interest in the role of objects and the material world in International Relations and its associated fields. Sometimes this is called the 'new materialisms' (Coole & Frost, 2010). You can see it in the interest in feminist theorizing about the role of the body. You can see it in environmental politics in the sensitivity to the limits of the planet, in sustaining the type of global system that we have. You can see it in those inspired by the Marxist tradition and a renewed interest in global inequality and the material inequalities that are associated with that. This interest in technology and in Actor Network Theory and similar approaches is part of an exciting recognition of the material properties of the global system that have always spanned borders but have sometimes been overlooked, especially in the state-centric period where the international spaces were seen as relatively empty. Today it's very evident that there are no empty spaces—that they are filled with objects, non-humans, and human activity of all types.

Malcolm: I want to thank you for your time Tony, it's been a real pleasure.
Tony: Thanks a lot Malcolm.

NOTE

1. On this see: https://www.commerce.senate.gov/public/index.cfm/
pressreleases?ID=2BA05D9F-EAAD-4043-B195-B8D91FF690B4.

WORKS REFERENCED

Callon, M. (1991). Techno-economic networks and irreversibility. In J. Law
(Ed.), *A sociology of monsters: Essays on power, technology and domination* (pp.
132–164). London and New York: Routledge.

Campbell-Verduyn, M., Goguen, M., & Porter, T. (2017). Big Data and algorithmic governance: The case of financial practices. *New Political Economy,*
22(2), 219–236.

Coole, D., & Frost, S. (Eds.). (2010). *New materialisms: Ontology, agency and*
politics. Durham: Duke University Press.

Dobusch, L. (2012). Algorithm regulation #2: Negotiating Google
search. *Governance Across Borders,* August 11, 2013, http://govern-
ancexborders.com/2012/08/11/algorithm-regulation-2-negotiating-
google-search/#more-2836.

Karppi, T., & Crawford, K. (2016). Social media, financial algorithms and the
hack crash. *Theory, Culture & Society, 33*(1), 73–92.

Latour, B. (2003). The promises of constructivism. In D. Ihde & E. Selinger
(Eds.), *Chasing technoscience: Matrix for materiality* (pp. 27–46).
Bloomington and Indianapolis: Indiana University Press.

Latour, B. (2004). *Politics of nature: How to bring the sciences into democracy.*
Cambridge, MA: Harvard University Press.

CHAPTER 9

From Nuclear Weapons to Cyber Security: Breaking Boundaries

Madeline Carr in conversation with Joseph Nye

Abstract Joseph Nye has been working at the intersection of technology and international relations for half a century. In this interview, he recounts some of his early experiences of working with academics who were willing and able to collaborate across disciplines. With great narrative instincts, Nye explains how he engaged with scientists and the technical community to understand and interpret the foreign policy dimensions of developments in both nuclear science and ICTs. He also speaks eloquently of his belief in the value of university departments able to foster the innovation and curiosity of the next generation of IR scholars working to understand the implications of technology for international relations. Nye's account is fascinating not only because he provides great insight into how digital technologies can be understood in IR but because in doing so, he traces a path through his extraordinary career.

M. Carr (✉)
University College London, London, UK

J. Nye
Harvard's Kennedy School, Cambridge, MA, USA

Keywords Digital technology · Cyber security · Foreign policy · Internet · Interdisciplinarity

Until relatively recently, cybersecurity had not attracted a great deal of attention from the discipline of IR—certainly not the same as some other large technological developments like nuclear technology. Given the scale and scope of challenges that digital technologies pose to central concepts like war, international law, diplomacy, global governance and international political economy, this has been somewhat perplexing for those of us who have been studying it for the last decade or more. One scholar from the canon of American IR has consistently engaged with the 'information revolution' as it has unfolded. As far back as 1998, Joseph Nye co-authored (with Robert Keohane) an article on 'Power and Interdependence in the Information Age'. Nye has continued to work towards understanding what cybersecurity means for international relations, particularly through questioning the implications for power (2010), and, more recently, deterrence (2017). By his own admission, Nye is not particularly focused on the technology itself—rather, he brings decades of experience of thinking about how new technologies impact on IR, where he can see continuity, and where he observes change. This recognition that technology is not new to IR was exactly the impetus for this volume. Nye's account is fascinating not only because he provides great insight into how digital technologies can be understood in IR but because in doing so, he traces a path through his extraordinary career.

What is particularly useful here is Nye's explanation of how he engaged with scientists and the technical community to understand and interpret the foreign policy dimensions of developments in both nuclear science and ICTs. This, he suggests, requires a level of translation skills that will be familiar to many STAIR scholars who find themselves communicating to multiple audiences. Nye also recounts some of his early experience of working with academics who were willing and able to collaborate across disciplines and his belief in the value of university departments able to foster the innovation and curiosity of the next generation of IR scholars.

Madeline: Joe, looking back on your long career, how do you feel technology has been present in your work? What kind of an influence has it had on your thinking about IR?

Joe: Well, I think technology has been important in the way that I approach IR, but I would say particularly important since the early 70s when Bob Keohane and I did our work on 'Transnational Relations and World Politics' (1972) in which we were trying to get away from the state centric view of world politics. We realized that there were many actors that had been empowered to play a greater role in world politics because of technological developments, particularly information technology. They weren't totally new, obviously. For example, multinational corporations had been around for a century, but the ability to have an integrated global strategy was greatly enhanced by information and communications technology. Similarly, terrorism had been around since the nineteenth century. But this was greatly advanced by the changes in technology. So as Bob and I were editing this volume which was designed to get people to broaden their horizons beyond a state centric approach, I think we were very much aware of the role that technology played to enable or empower many of these non-state actors.

Transnational flows got me interested in the trade in nuclear materials and the proliferation of nuclear weapons. I went into the State Department as a political appointee in 1977 to run President Carter's policy on non-proliferation of nuclear weapons. That obviously had a heavy state component about it but it also had a non-state component. More important though, was the way in which the technology—nuclear technology—was beginning to enable states that previously had not been in a powerful position. During this time, the Carter administration stopped the reprocessing plant that we had at Barnwell, SC and we also halted the Clinch River Breeder reactor which would have produced great quantities of plutonium. We tried to convince other countries to halt—not all nuclear activity—but to make these distinctions within nuclear technology. So that took a fair amount of learning about the technology, trying to understand what distinctions you could make, how you could justify them and where you could develop consensus amongst states about them.

Madeline: *Sounds very similar to what's been happening in this room today [at a workshop on cyber norms]. How did you bridge that policy/technology gap?*

Joe: Oh, very much so! It was interesting—I had freshman physics but I was certainly not a physicist. I spent time at the National Laboratories, I had colleagues at Harvard who worked in nuclear engineering who acted as advisors, I hired a nuclear physicist to be on my staff in the State Department so I'd have somebody on my right hand who could advise me. I think we had a certain degree of success and at least I got a State Department Award for the work so I can't have messed up too much on the technology![1]

When I finished that two years leave from Harvard, I taught a seminar on nuclear ethics and then published a book with that title—*Nuclear Ethics* (1986)—trying to say, where do we make these distinctions [about who should and should not have access to nuclear technology] and how do we justify them ethically as well as politically? In the early 80s when Reagan was elected, there was great concern about nuclear war. Derek Bok, the Harvard President, came to five of us who were on the faculty and suggested that we write a book about how we should deal with nuclear weapons. So we collaborated on a book called *Living with Nuclear Weapons* (1983)—so, how do you *live* with nuclear weapons? And then at the Kennedy School, Graham Allison, Al Carnesale and I followed that up with a project supported by the Carnegie Foundation looking at the practical steps one could take to reduce the probabilities of nuclear war which we called the 'Avoiding Nuclear War' project.

After that, I became interested in the questions of what was going to happen to American power at the end of the Cold War. And again, one of the questions that came up was how could we think about American power and what role did technology play? In the process of thinking about that, I also became interested in the role of Japan. There was a great deal of concern about whether Japan was going to surpass America technologically. I then went into the Clinton administration, first as Chairman of the National Intelligence Council and then as Assistant Secretary of Defence. One of the things we did was to develop a strategy for East Asia and the questions were, 'how was technology strengthening China and Japan? What was going to be the balance between them? And what was the American role?' And so again, technology was a major factor for that assessment. By the time I came back from government to be Dean of the Kennedy School, I had become intrigued by the effect of the Internet which was beginning to have a visible impact on politics and society.

Madeline: *What year would you say that was, Joe?*

Joe: I came back in 1996 as Dean and we had a project called the '21st Century Project' that looked at what would be the major issues for government to deal with in the twenty-first century. One of them was how we were going to cope with technology and the Internet. Elaine Kamarck and I—Elaine had worked for Al Gore and she was a lecturer in the School—she and I edited a book called *Democracy.com* which was asking questions about the Internet and its effects on domestic policies.

Madeline: *In the mid 1990s? That was very early.*

Joe: Yes, I think that book was probably published in about 1999 or 2000. And then I had begun to see the impact of the Internet. We hired some young people on that but I didn't really become deeply involved in cyber issues until about 2008. I mean, I still kept my interest in Internet issues but, Bill Studeman, a former director of the NSA who is here today at this conference, came to visit me and said 'you have to pay much more attention to cyber politics and cyber issues' and I said 'well, Bill, I'm too old for that – it's too late' [laughter] and he said 'no, you don't need to become a cyber expert but you're a foreign policy expert and this is going to be one of the major issues in foreign policy'. I took him seriously and started trying to educate myself on cyber issues and cyber politics and one thing led to another on that. Now I've published several articles including 'Nuclear Lessons for Cyber Security' (2011) which was an effort to say ok, when you have a new technology like nuclear weapons, how do states learn? In the 80s, I'd written an article called 'Nuclear Learning' (1987) which was trying to trace how attitudes developed and how states began to develop principles for approaching nuclear weapons. I thought 'are there any lessons from that for cyber'?

Now the two technologies are totally different so the direct lessons made no sense but there was a meta lesson which was 'how do states learn about a massive disruptive new technology?' Cyber was becoming that new technology. The focus for me became how we *thought* about an issue—how long it takes states to learn about it, to come to agreements—and there, we had some interesting analogies. In nuclear weapons the first agreements don't come till about 15 years after—I mean, Hiroshima was 1945 obviously and the Limited Test Ban Treaty is 1963 and the NPT is 1968 so it's a decade to a decade and half before states learn the principles of cooperation. And it's interesting really that the antecedents of the Internet had been around since the early 70s but cyber really doesn't take off as a substrate of the international economy and politics until really the late 1990s. And if you project forward 15–20 years after that…

Madeline: More or less where we are now…

Joe: Yes. And the question is, what do we know about how states can learn, develop cooperation and develop norms in that area. I was invited to join the Bildt Commission [Global Commission on Internet Governance] and I wrote a paper for them about regime complex for cyber activity. I taught some courses for the Kennedy School on regimes for cyber and then wrote a piece that has just been published on dissuasion and deterrence in cyberspace which is in this current issue of *International Security* (2017). But the common theme through this was not that I was going to become a cyber expert but that maybe I could be an interpreter between traditional IR theory and this new technological area.

Madeline: *As you had done in the past with nuclear technology.*

Joe: Similar to that. After the Bildt Commission ended, the Dutch government announced at the Munich Security Conference (2017) that they were creating a new Global Commission on Stability in Cyberspace (GCSC) that Marina Kalljurand (former Estonian Foreign Minister) is chairing. They asked if I would join that and I agreed. We had a panel on this at the Munich Conference and somebody there asked me what I expected to be able to do on the GCSC. I said I will never be a cyber native. My granddaughters will be but I won't. I will always be a cyber immigrant and I will always speak with a heavy accent but even immigrants with accents can act as interpreters and translators between cyber technology and policy. And so that's what I see my goal as. It's to try to be a link between an area which has been developed rapidly by very brilliant technologists who don't necessarily have a lot of experience of international politics and somebody who has had a career working in international relations, both theory and practice, but who can then serve as a kind of a link in that area.

Madeline: *You said that you had basic physics from college which might have helped you while working on non-proliferation. There can be a certain barrier to engaging with the politics of digital technology for those who lack technical literacy. Do you see this differently or was there the same kind of resistance from nuclear scientists?*

Joe: Oh, I think there's a resistance. The people who come up through the technology—who are true experts—are resistant to people who are poaching their territory. When I went to Oakridge, Tennessee to announce at a conference of nuclear engineers that the Carter administration was stopping the Clinch River Breeder Reactor, one of the engineers got up and said 'what right do you have to interfere in something which for us, is sacrosanct, which is that the nuclear fuel cycle should progress from uranium to plutonium?' It's almost like a priesthood for them. I mean they had believed that this was fore-ordained by physics.

In fact, there are some critical human choices which are based on economics and politics but they didn't want to hear those because they interfered with what they thought was the natural order. So, there is a natural resistance by people who've spent their lifetime in an area and who believe that if you've approached it from the point of view of an engineer, it's quite different from approaching it from the point of view of an economist or a political scientist. I mean for an engineer, getting the most energy out of an atom is crucial and therefore, you don't want to waste any atoms by putting uranium through a reactor once and then burying the residue in the ground. From an economist's point of view, it might be better actually, to mine new neutrons from underground and to bury the residue. And from a political scientist's point of view, it might be far less dangerous to leave these atoms and neutrons buried underground rather than make them readily available in a plutonium trade where it can be diverted by terrorists or others more easily.

So, each discipline brings a different perspective to a question. For the nuclear engineer, it's energy and efficiency and using nature's bounty in the most efficient way. For the economist, it's using nature's bounty in the way that makes most economic sense. And for the political scientist, it's both the latter and the question of security and making sure that we don't blow ourselves up.

Madeline: How do we get out of our disciplinary silos to build a more sophisticated dialogue amongst ourselves?

Joe: Well, I was struck that many of these young people who were coming along as pre-doc and post–doc students were pretty good at doing that. They were already thinking across silos. The Belfer Center for Science and International Affairs was founded by a chemist, Paul Doty who got very involved with arms control issues. He essentially wanted to provide a welcoming environment for people who wanted to cross boundaries. Albert Carnesale, who was the Assistant Director of the Belfer Center, was trained in nuclear engineering. After I came back from government, I succeeded as Director and Ash Carter, whose training was in physics, became the Assistant Director with me (and then of course, he went on to become the Secretary for Defence and subsequently returned to direct the center). You can create institutions where people talk to each other across disciplinary bounds and then you can also bring along young people who are better at that.

Madeline: Harvard has obviously been very good at doing that but what do you think when you look around at other universities? Do you think that they're doing what they can to facilitate that for students? Because typically, when students come to do a degree program, at least in the UK, they're locked within their discipline and it can be difficult to arrange for them to go and take some modules in Computer Science, for example.

Joe: Well, I think that's true and it's going to require universities to think about how you get people to be conversant across disciplines. It's difficult to get people to rise in more than one discipline because peer review of articles and peer review of appointments means that you have to excel in one discipline. But you can still have people who excel in one discipline but can be conversant with, or literate in, another discipline. I think that's probably what we try to aim for.

Madeline: Why do you think, Joe, that over the last 25 years there has been so little happening in IR in terms of digital technologies. There's been a little bit but you might expect that after 25 years, we would have seen more in our discipline.

Joe: Well, I think until recently it's been difficult for young scholars to convince the mainstream of the discipline that this was a major issue and so if you were looking at journal articles of the early 2000s, you would note that very few issues of *International Security* or *International Organization* or *American Political Science Review* would pay any attention to these issues. What's interesting though is the way that academic fashion, with a lag, follows political life. The Director of National Intelligence in the US provides an annual report to the Congress on major threats and in 2007 or 2008, I think, the issue of cyber didn't appear in the list, or if it did, it was way down the list. By 2015, it was top of the list. We have 2.5 billion people connected to the internet today and we had fewer than 100K around 1999. That exponential growth has affected the dependence on the Internet and with increased dependence comes vulnerability and with vulnerability comes security problems and it suddenly becomes more interesting. Now you're seeing a good number of books and articles on cyber and more realization by [university] departments that they've got to cover cyber where 10 years ago or even 5 years ago, they saw this as sort of a side track. They couldn't quite see why it was relevant.

Madeline: Last question: what do you see as the most significant challenges and opportunities for IR in the context of ongoing technological change?

Joe: Well I think there are going to be a large number of technologies which are going to raise fascinating questions. I mean, some people say that bio-technology will be to the twenty-first century what nuclear technology was to the twentieth century—that's a bit too simplistic but if you are looking at things which can lead to genetic engineering so that you can create different types of human beings and extend the lives of human beings considerably, well, what does that mean? Should any laboratory in any country be allowed to create any kind of human hybrid? Should it be stopped?

And then there are the issues that are already with us in terms of climate change. Some people say that it's going to require climate engineering. My colleague at Harvard in the Kennedy School, David Keith, is looking hard at questions of whether you could spray aerosols into the atmosphere to block sunlight to counter the effects of carbon dioxide. Well, if you did then that might prevent sea level rises that might save the Maldives and Bangladesh and whatever, but who decides that and how do you control that and what are the side effects and unintended consequences? What's it going to do to fish when all that stuff precipitates out? And how do we organize to make judgements about that? Those are things that are going to be of tremendous importance to IR—they already are but I would suspect increasingly so.

And another area of course, is artificial intelligence and machine learning—that question of the smart robot. The issues relating to AI and machine learning are going to be extremely important. So, I think there is going to be no shortage of need for International Relations departments to have people who are able to at least be literate or conversant with these technological problems because I think they're going to be increasingly important to politics and to our lives. And then you have to ask whether the discipline is adjusting adequately to prepare for that and I think there are considerable lags. Many of these things are still regarded as fringes or fads but I think a good Department chair or good University President or Vice Chancellor will start thinking on these things.

Madeline: Where do you think the disciplinary lags are? What is it about IR? Because on the one hand you could say that there is a lot of useful thinking in IR that could help us understand the implications of new technologies. And then sometimes when you put the two together, you think there are some problems with IR that don't really help us get to the heart of the problems.

Joe: Well the kinds of debates that we have about realism vs neo-liberalism vs constructivism—I don't think they help very much. No one of those has the answers and the question is how can you take the insights from all three of these broad theoretical approaches and use them to understand things. I mean, realists will say 'what about China and Russia and the dangers of cyber war and cyber conflict?' Well, that's a valid insight. Neoliberals will say 'what about institutions and the development of norms?' That's a valid insight. Constructivists will say 'what about epistemic communities and what about broad public thought in terms of rearranging the way we think about what a problem is?' It's not going to be one or the other of these—it's going to be all three.

Madeline: Breaking out of these boundaries...

Joe: Right. But our traditional debates, as though there are three approaches, I think are a hindrance more than they help us. That's why I've taken to calling myself a 'liberal realist' or maybe I should call myself a 'liberal realist constructivist' [laughter].

Madeline: Or a constructed liberal realist! Thank you very much, Joe. You've been very generous with your time and it's been a fascinating insight into your thinking.

NOTE

1. Nye received a State Department Distinguished Honor Award in 1979.

WORKS REFERENCED

Carnesale, A., Doty, P., Hoffmann, S., Huntington, S. P., Nye, J. S., Jr., & Sagan, S. D. (1983). *Living with nuclear weapons.* Cambridge: Harvard University Press.

Kamarck, E. C., & Nye, J. S., Jr. (1999). *Democracy.com?: Governance in a networked world.* Hollis: Hollis Publishing Company.

Keohane, R. O., & Nye, J. S., Jr. (Eds.). (1972). *Transnational relations and world politics: An introduction.* Cambridge: Harvard University Press.

Keohane, R. O., & Nye, J. S., Jr. (1998). Power and interdependence in the information age. *Foreign Affairs, 77*(5).

Nye, J. S., Jr. (1986). *Nuclear ethics.* London: Collier Macmillan.

Nye, J. S., Jr. (1987). Nuclear learning and U.S.-Soviet security regimes. *International Organization, 41*(3), 371–402.

Nye, J. S., Jr. (2010). *Cyber power.* Cambridge: Harvard Kennedy School, Belfer Center for Science and International Affairs.

Nye, J. S., Jr. (2011). Nuclear lessons for cyber security? *Strategic Studies Quarterly, 5*(4), 18–38.

Nye, J. S., Jr. (2017). Deterrence and dissuasion in cyberspace. *International Security, 41*(3), 44–71.

CHAPTER 10

Technologies of Violence

Myriam Dunn Cavelty and Jonas Hagmann in conversation with Keith Krause

Abstract In this written transcript of a conversation with Keith Krause, Professor at the Graduate Institute of International and Development Studies (IHEID) in Geneva, Switzerland, Director of its Centre on Conflict, Development and Peacebuilding (CCDP), and—until 2016—Programme Director of the Small Arms Survey, central aspects of the contemporary use and study of technologies of violence are highlighted. The ways in which technology informed Keith's research and how it altered the character of international violence are addressed, before the conversation moved to how technology affects patterns of knowledge creation and dissemination in the IR discipline more generally.

M. Dunn Cavelty (✉)
Center for Security Studies, ETH Zürich, Zürich, Switzerland

J. Hagmann
Institute of Science, Technology and Policy, ETH Zürich, Zürich, Switzerland

K. Krause
Department of International Relations/Political Science, Graduate Institute of International and Development Studies (IHEID), Geneva, Switzerland

97
C. Kaltofen et al. (eds.), *Technologies of International Relations*,
https://doi.org/10.1007/978-3-319-97418-7_10

Keywords Technology · Violence · Small arms · Critical security
studies · Public policy · Digitalization

Few things may impress observers as much as the persistence of violence
in international politics. Yet, uses of violence in world politics are considerably more diverse in practice than what the classic International Relations
(IR) focus on military national defence suggests. One of the leading
international experts on contemporary armed violence is Keith Krause.
Professor at the *Graduate Institute of International and Development
Studies* in Geneva and Director of its *Centre on Conflict, Development and
Peacebuilding*, Keith Krause had been at the fore in questioning the performativity and utility of conventional IR security concepts,[1] and a crucial
voice in the establishment of critical security studies.[2] His wide corpus of
published work includes empirical studies and analytical reflections alike,
on international arms trade,[3] human security,[4] post-conflict state- and
peacebuilding,[5] and small arms[6] in particular. In 1999, the latter focus
resulted in his funding of *Small Arms Survey*. The internationally recognised research centre NGO, of which Keith was Programme Director
until 2016, produces annual volumes on small arms proliferation, stockpiles, transfers, misuse and effects, including field- and issue-based studies.
Today, *Small Arms Survey*[7] is the main global source of information and
analysis for international public policy on small arms issues.

   With his critical analyses and empirical knowledge of the changing
character of armed violence, years of professional experience in academia
and impressive ability to inform policy-making by reflexive scholarship, we
considered Keith Krause an ideal candidate for a conversation about the
(changing) role of new and old technologies, whether in practical world
politics or the IR discipline itself. In our 30-minute conversation, which
took place by Skype in early May 2016, we directed the discussion to what
we felt are central aspects of the contemporary use and study of technologies of violence—not just the ways in which technology informed Keith's
own research, or how it altered the character of international violence, but
also how it affects patterns of knowledge creation and dissemination in the
IR discipline more generally. The conversation was transcribed with the
help of two Research Assistants, Patrice Robin and Darius Farman, and
then edited for style and consistency by us interviewers.

***

Question: Thank you very much for your time, Keith. Let's get right into it: To clarify the context and concepts, what are technologies for you?

Keith: There is a broad and a narrow sense of technology. In my own work, I mostly focus on technologies in the narrow sense, understood as actual instruments, artefacts; usually weapons, instruments of violence, and things we use to manipulate or change the world. There is also a much broader sense and I will touch upon that later, where we can think of technologies as form of knowledge or instruments in different social relations.

Question: What role do technologies play in your empirical work and to what extent did they influence your conceptual or analytical perspective?

Keith: Well... a long time ago in a galaxy far, far away (laughs), I worked on the political economy of arms production and transfers. One important aspect of that was to look at the process of innovation, diffusion and application of technologies. This was a large, long-term state-centric historical analysis of arms diffusion that resulted in a book with Cambridge.[8] What was most fascinating to me was the relationship between the pace of innovation, the attempts to regulate or control diffusion, and the relatively systematic obstacles to the uptake of various innovations by what we can call "second tier actors". It shed some important light on two things: (1) that the process of diffusion is socially conditioned and not automatic and (2) that innovation and diffusion work together. While diffusion may accelerate if innovation is happening more rapidly, the notion that there is some inevitable process of levelling or catching up is not necessarily true.

In terms of more recent work, I started focusing particularly on small arms and light weapons, work that lead to the establishment of the *Small Arms Survey*. Here, we have a tighter, less conceptual focus. The idea was to bring the instruments of violence back into various disciplines that studied armed violence, ranging from war and conflict to localized inter-personal violence. The other two communities that focused on the issue – principally public health and criminology – followed a strategy of "we'll try and assess what we can measure". They had useful data on demographic risk factors, age-related issues, spatial risk factors, or things such as alcohol abuse and violence – but they never focused on the instruments of violence, because they had no data. Perhaps they assumed that instruments of violence can be taken as a constant or that they exist everywhere and people just have them, which is obviously not true. So it was interesting to try and highlight the broader social and political context of instruments and technologies of violence for research communities like public health and criminology.

Question: You have already touched upon the effect of technologies on security and politics a bit, but can you say something more specific about the relationship between technology and power?

Keith: Okay, that is a wide-ranging question and I think that there are a couple ways of cutting into this. If we take technology in a narrow sense as the artefacts and the instruments of violence, then it seems fairly obvious that there is more lethal technology of greater sophistication more widely diffused and more easily accessible than at any time in the past. That certainly causes us to rethink what we mean when we talk about the monopoly of the legitimate use of violence, because now the accent is placed much more on the legitimacy rather than the practicality of this monopoly. That raises issues about state capacity, especially about the provision of public order and security as a public good. I would say that power in this narrow (internal) sense has diffused. You can see that in the rise and the increased sophistication of non-state armed actors and the way in which they can operate in multiple contexts.

This is a little bit simplistic, however, because it ignores technology in the larger sense, which is the broader social context in which these means are embedded. The question is: do the other factors that we would associate with technology – like acquisition, use, or the ability to transform local social relations – also operate to diffuse power throughout the system, not just from states to citizens, but also from strong states to peripheral client or weak states? It is far less clear whether or not that kind of diffusion is occurring. I do not have a crystal ball that can arbitrate between the pure diffusion advocates, who say: "1000 flowers will bloom and everybody will be armed and everybody will be safe", and the others who say that there is still a really powerful hierarchy, because the basis for ultimate innovation and the technologies of social control are not just about weapons but also about surveillance and monitoring, and will ultimately result in consolidation of state power rather than a weakening of it.

Question: Digital technologies are recurrently said to affect world politics particularly strongly. In your view, has the advent of the digital age changed the nature or importance of other forms technologies of violence and insecurity?

Keith: If we take digital in its broader sense, as communication and transmission of ideas and intangibles, then it is very clear that there are many forms of institutional and social mimicry and isomorphism, which is digitally transmitted. Some of this is humorous and anecdotal, such as the way in which warlords and commanders in armed groups in West Africa adopt particular styles of sunglasses, certain kind of poses or the nicknames they give each other that are gestures towards Hollywood films. Whether or not this actually generates any local power and legitimacy is hard to assess, however, because it is speaking in a sort of global sphere to a set of symbols that they wish to demonstrate their mastery of. On the other hand, certain groups like the Lord's Resistance Army still make a lot of their mileage out of appeals to magic and witchcraft, which are – shall we say – slightly antithetical to the way we tend to think about the digital age, although I think there is room for both in the digital sphere. And this is part of the process of asking and understanding what is transmitted along with weapons and know-how.

There is one other important part of the digital age, which is how knowledge about using artefacts can be acquired. I am sure we all watch YouTube videos to try and figure out how our coffee machine or printer works. Well, you can kind those kinds out there for how to operate some types of weapon system – how to take apart, clean and rebuild your AK47 and various other things for example – that is definitely a levelling device. We also know that a YouTube video is not a perfect training tool, but it does mean that people are able to acquire knowledge about the use of artefacts that is a bit different than in the past and I think more "democratic".

Question: *That brings us to our next question about the challenges and opportunities. If you think about technological change more broadly but also in your particular field of small arms as instruments of violence, what do you see as the most significant challenges and opportunities for world politics in the broader sense?*

Keith: Do you mean world politics as the discipline of International Relations?

Question: *We will come to the discipline a bit later, so maybe in terms of international politics as a political practice first.*

Keith: Sure. If we think about the practice, what is interesting is the disconnect between the diffusion of low-tech weapons versus a continued high-tech fetishism, which is clearly manifest in the United States policy and military circles, but also in other countries. Basically, it is about the idea that there is a technological fix to their security challenges. A good example is the widespread use of drones as a way of presuming that there can be bloodless war, at least on the side of the West. This is in sharp contrast to the opposite view dating back to the first studies on counter-insurgency, which says that an asymmetric conflict is one in which local actors are very adaptive to their technological disadvantages and are able to use what is at hand, like simple GPS tracking devices to hide arms and caches, which allows them to decentralize their operations, or widespread use of encrypted satellite phones and other forms of communication that makes them hard to track. These kinds of things "democratized" the advantage that the leading states often think they possess. This is not a new story, but it is taking on a particularly strong dimension in the endless war against terror. The main question though is: "is who is empowered and who is disempowered by these technologies within and between states?". The best way to understand this is not through a traditional IR balance of power approach, where you can estimate the importance of technology either in the size of your armed forces or the number of your soldiers, but rather through the much more diffused and socially embedded understanding of who is able to manipulate technology, what is available to them to achieve their ends and how this might cause grave insecurity for people around them. That is a much more widespread type of diffusion, which you cannot capture if you just look at the balance of power in terms of arms spending and such.

Question: We now want to shift the conversation away from politics as practice and towards the discipline of IR. To what extent have technologies changed the ways in which IR scholars approach key concepts, like for example security, power and order?

Keith: To approach this, I want to mainly focus on the technologies of knowledge creation first. A key lesson from the last fifteen years for me is the rise of evidence-based knowledge into a particular position of power. Evidence-based knowledge is based on techniques and models from (mainly) public health, and you can really see the language of randomised control trials starting to sneak into International Politics, in terms of natural experiments or experimental approaches. It creates a huge disconnect between the forms of knowledge that we try to create in either International Politics or in any critical approaches and much more hard-nosed "scientistic" approach. These have really powerful built-in biases, in particular, they are based on a certain individualist ontology and emphasise agency, but do not have much to say about structures in the long-term and historical changes. Even communities that straddle this, mainly the criminologists, are divided: some look to one side and see public health people who have these fabulous tools where they go in and do interventions on violence prevention in Kigali among sixteen families and separate them into control and treatment groups, and the others look in the other direction and say: "Well, what are the broader structural conditions that could give rise to poverty and inequality, or issues of how identity is created and reinforced?" Almost none of these are going to be captured by the technologies of knowledge creation that are privileged in the model of evidence-based knowledge. The problem is that evidence-based knowledge is really very powerful when you put it out in the public sphere. For policy-makers, if you do not manage to put evidence-based knowledge into your presentation, they think you are just telling stories. That is a real challenge because advocating and arguing for a story-telling approach runs up against this other understanding of the technologies of knowledge creation.

So, what does that say about security, power and order? I think in terms of how we study these, the concepts are increasingly defined by the measurability of them. That is not new per se, but, as the earlier example I gave about criminologists and people studying violence shows: what gets measured gets done, what gets measured gets studied. Even the way people define violence and security tends very much to be tilted towards what it is they can acquire "reliable knowledge" about. How many victims are there or how many people have injuries that we can collect data on from hospitals, as opposed to what are the psychosocial dimensions of trauma of a community, which is not measurable in the same way.

Question: You already touched on the dissemination of knowledge when mentioning evidence-based knowledge production and the communication of policy-makers. What about other forms of knowledge diffusion, such as in classrooms? To what extent have technologies affected teaching practices in the discipline?

Keith: Sitting in Geneva and teaching graduate students only gives one very narrow perspective on what this means. I am going to flip it around and say what we see increasingly is an emphasis on matching teaching with training, and training is usually understood to be training in techniques. We have *techne*, techniques, technologies, certain ways of generating and acquiring knowledge, so you may teach a subject but you train a method. There is an increasing emphasis on equipping students with the appropriate *tools* –and that is not an accidental term– because they do not usually mean those in theoretical terms, to go out and study the world. I think that is affecting teaching. Some of this is not bad. I do think that everybody needs to have some basic understanding of how statistics can be used and abused, because it happens to us all the time. But there is just an inevitable tilt towards certain forms of knowledge here that again leave out essentially social, contextual, structural, local knowledge; the sorts that James C. Scott would call the *metis*. Knowledge is embedded in particular times and places. It is not abstract and general – and always measurable.

Question: If we pursue this conversation a little bit further and look at how careers are being made and developed in the discipline. Do you see effects of new technologies on the ways in which scholars progress through academia?

Keith: I think so. You could argue that the focus on measurable metrics in citation counts and journal rankings is of course a technology of comparing knowledge and it has one singular advantage, which is that is relatively transparent. But everybody who studies this in any serious way knows that behind that is a set of power relations that both structure the discipline and individual career chances. There was a brilliant study by a Harvard-trained economist[9] – who is a woman – who pointed out that co-authored papers by women were discounted systematically in tenure decisions, that is to say if there was a man and a woman co-authoring, the woman's contribution was devalued. And so, the disclaimer is: do not co-publish if you are a woman. This turns out not to be the case in some other disciplines, so we can interrogate our economist colleagues about this, but behind these technologies there are other older, informal forms of disciplining that are often used instrumentally in order to maintain whatever academic norms are considered to be appropriate. I think that is an old story, it has not changed much. The real question is has it become worse or better? I do not think we have a really good handle on that.

Question: If we move beyond academia, have new technologies made it easier or more difficult for academia to connect with policy-makers?

Keith: I essentially work with ten or twelve governments focusing sometimes very narrowly on issues of how we can prevent the diffusion of arms to illicit armed groups or more broadly how can we design appropriate evidence-based violence reduction interventions. What people are looking for in the policy world is digestible knowledge that makes certain gestures towards its scientific reliability. They want the findings and then they want to know that the method is valid, in some important sense. That is both good and bad. In our work, one of the things that we pushed very hard in the *Small Arms Survey* was to improve the quality of analysis, so that people did not think that a research paper was simply hiring a consultant and having them interview 20 people in a capital city about what they thought was going on. There are technologies of knowledge, like focus groups and structured interviews, that I think are important. So, making sure the knowledge that was created had some more robust foundation was a good thing. But the question is how far should that go? Do you end with randomised control trials when you start down that path, considering that they are the only gold standard for some? I spoke to a research advisor for the Foreign and Commonwealth Office about this last week, he is a former academic, and he laughed and simply said: "Well, you know, I had to fight really hard to get people to abandon this idea that the only reliable knowledge was that collected in these kinds of control trials or experiments" so there was a little bit of room for something that was contextual. So, I do think that we went down that path hopefully and may have ended up not realising some of the unanticipated consequences of doing so.

Question: Do you think there is enough or not enough focus on technologies in IR?

Keith: I would say there is not enough focus on the way in which technological artefacts are socially embedded, how they gain both their power as symbols, how they gain their ability to be used – in both the narrow sense of people understanding how to make it work, but also to use them as tools of gaining your legitimacy. I think without understanding the embeddedness in the social context, technological artefacts are usually just sticks and stone, or bits of metal. The focus on bringing in technology if it leaves behind any understanding of the social is not going to be very helpful. There is no doubt that we can see some of that with people who are wowed by the digital age and the potential of new technologies to empower or transform lives of people in far-flung places – without any reflection on how it actually could "work" for those people. As long as we can bring the social in, with the technological, we might gain some greater insights. But if it becomes fetishism of the new and the instrument itself, then we are unlikely to make any progress.

NOTES

1. Keith Krause. (1991). Military statecraft: Power and influence in Soviet and American arms transfer relationships. *International Studies Quarterly* 35(3): 313–336; Keith Krause & Michael C. Williams. (1996). Broadening the agenda of security studies: Politics and methods. *Mershon International Studies Review* 40(2): 229–254.
2. Keith Krause & Michael C. Williams. (1997). *Critical security studies: Concepts and cases.* Minneapolis: University of Minnesota Press; Keith Krause. (1998). Critical theory and security studies: The research programme of 'critical security studies'. *Cooperation and Conflict* 33(3): 298–333.
3. Keith Krause. (1992). *Arms and the state: Patterns of military production and trade.* Cambridge: Cambridge University Press.
4. Keith Krause. (2002). Une approche critique de la sécurité humaine. In *La sécurité humaine*, Jean-François Rioux (Ed.). Paris: L'Harmattan, 73–98; Keith Krause. (2004). The key to a powerful agenda, if properly delimited. *Security Dialogue* 5(3): 367–368.

5. Keith Krause & Oliver Jütersonke. (2005). Peace, security and development in post-conflict environments. *Security Dialogue 36*(4): 447–462; Keith Krause. (2012). Hybrid violence: Locating the use of force in post-conflict settings. *Global Governance 18*(1): 39–56.
6. Keith Krause. (2001–2011). *Small Arms Survey* (Eleven Editions). Cambridge: Cambridge University Press.
7. See http://www.smallarmssurvey.org/.
8. Keith Krause. (1992). *Arms and the state: Patterns of military production and trade.* Cambridge: Cambridge University Press.
9. Note by the editors, this is Heather Sarsons. (2015). Gender Differences in Recognition for Group Work, Working Paper, available at http://scholar.harvard.edu/sarsons/publications/note-gender-differences-recognition-group-work.

CHAPTER 11

Postinternationalism on Technology, Change and Continuity

Stefan Fritsch in conversation
with Yale H. Ferguson

Abstract Postinternationalism provides a flexible theoretical perspective to explore technology in its manifold variations and diverse impacts on global politics, economics, security and identity. Postinternationalism embraces a broad and non-deterministic conception of technology. This epistemological and ontological position enables its representatives to analyse technology's ambivalent and multifaceted impact on the global system more effectively than other IR theories, which often emphasize very narrowly defined factors as pivotal for their explanatory models. In this context, Postinternationalism perceives technology as a man-made enabler of diverse and often seemingly contradictory processes ranging from the integration of polities and transnational interaction processes to

S. Fritsch (✉)
Department of Political Science,
Bowling Green State University, Bowling Green, OH, USA

Y. H. Ferguson
Rutgers University, Newark, NJ, USA

C. Kaltofen et al. (eds.), *Technologies of International Relations*,
https://doi.org/10.1007/978-3-319-97418-7_11

opposite processes resulting in the fragmentation of polities and even a potential slowdown of globalization. As such, Ferguson argues that technology must become a greater focus of scholarly exploration to further deepen our understanding of the driving forces behind change and continuity in global affairs.

Keywords Change · Continuity · Globalization · International relations · Internet · Polity · Postinternationalism · Technology

> *Stefan: Thank you so much Yale for agreeing to this short interview. We have prepared a couple of questions for this interview. I will jump right in. In connection to the overarching question of change and continuity and technology's role in it, how would you as a founding voice of postinternational IR theory, characterize the basic tenets of Postinternationalism?*
>
> Yale: Postinternationalism in IR theory is associated mainly with the work of our late friend and early mentor James N. Rosenau and the "polities" model of global politics that I and Richard W. Mansbach developed together. The Ferguson/Mansbach model was first advanced in our 1996 volume on *Polities: Authority, Identities, and Change*, then modified and expanded in 2004 in our book title *Remapping Global Politics: History's Revenge and Future Shock*, and most recently in our 2008 collection of essays *A World of Polities*.
>
> Postinternational theory arose out of dissatisfaction with the inadequacies and distortions inherent in traditional realist and neorealist theories, especially their narrow state-centric vision of the world. The postinternational view is that—although sovereign states and their "international" relations obviously remain important and are certainly going to remain so—state-centric theory simply misses most of the world of *global* politics that revolves around the behavior—not only of states—but also of nonstate actors within and especially across state boundaries. Post-internationalism focuses mainly on transnationalism and the dynamics of globalization. There is wide range of actors and polity types, for example, human individuals, tribes, perceived ethnicities, religious authorities and institutions, political parties, interest groups, corporations, bureaucracies and militaries (which may sometimes act independently from governments), cities, public and private international organizations, terrorist groups, organized crime, pirates, and so on and on. Each polity exhibits a degree of identity, and has elements of institutionalization and hierarchy (leaders and followers), and

capacity to mobilize its adherents. For their part, human identities are normally multiple, and only a few are mutually exclusive; what is important is not only the relative intensity of particular identities but also the fact that that intensity tends to wax and wane over time with changes in individual and group environments. Being Scottish has a different meaning, context and intensity at different stages of history. The Scotts are pretty much still, in many ways, what they always were. But they feel more Scottish at certain times than others. And this has to do with a variety of things that trigger this identity over others.

Postinternational theory emphasizes the dynamics of change in global affairs. Change is not unilinear (one-directional) in character nor is it necessarily progressive in a positive normative sense (it may be very bad change). Change may be very fast or so slow as to be almost undetectable, but evolution is nonetheless continuous. Postinternationalism stresses that change often takes the form of action/reaction or fission/fusion (integration-fragmentation or what Rosenau inelegantly termed "fragmegration"). Recall the frequently expressed observation that the world seems to be coming together and falling apart all at the same time. Sometimes, as at present, the accent is on discord and collapse (for instance, the European Union's current problems), but there is always some integration going on somewhere. For example, if I look at the current US presidential campaign, all candidates have critically remarked that free trade agreements with other countries hurt the US economy (which is a view I do not share). This could be considered as example for potential disintegration. Yet, at the same time, there are some very ambitious regional trade agreements being negotiated and one, the TPP, has been signed and is waiting (possibly in vain) for ratification. Mansbach and I also like Kal Holsti's threefold classification of change as being addition, or subtraction, or transformation. On the whole, postinternationalism likes to point out that contemporary globalizations differs from the past primarily because of the sheer volume and pace of present-day cross-border transactions. Everything seems to be happening much quicker. That said, we appear to be witnessing a slowdown of sorts now—and following so soon after the financial crisis—some aspects of globalization may actually be reversing. Our last book was *Globalization* and its subtitle was "The Return of Borders to a Borderless World?" (The question mark was significant.) We might be witnessing a period in which borders may be hardening, less goods are being traded, and the like. Or perhaps globalization is just changing in character—more digital and less physical goods, a process in which technology obviously plays an important role.

Stefan: How would you actually define technology from a post-international perspective?

Yale: As you know, the dictionary definition of technology is fairly limited, just something like the application of scientific knowledge for practical purposes. But I prefer the wider definition of technology being a realm of knowledge and activity that interrelates with life, society, and the environment—and of course includes, but is by no means limited to, industrial arts, engineering, applied science, and pure science. But technology exists in a broader context. It cannot be fully understood, appreciated, admired or criticized without that context.

Stefan: So you would say that you have a rather broad understanding of technology. Definitely not limited to the material artifacts as such.

Yale: Those are interesting in and of themselves, but it really limits the importance of technology and its broadest implications for human existence.

S.F.: In your view, which technologies are of particular relevance for a critical investigation in International Relations?

Yale: What is interesting here to me as a theorist is really having to come to grips with what we mean by the term International Relations. Different theorists have different views about what that is. This is a terrific question, because the answer so much depends upon one's perspective as to what is most important in IR. A classical realist would put most emphasis on the technology of warfare or on technology that enhances a national economy in a way that projects the influence of a particular country. It is not just your military hardware, but also your economic status and the amount of money you have available to invest in education, technology etc. A liberal would likely stress those technologies that enhance the potential for cooperation, interaction, exchange, and order. A constructivist would be especially interested in technologies that have the potential for shaping identities and perceptions. And so on. One of the beauties of postinternational theory is that you can have it all—any technology that potentially affects polities, authority, identities, transborder interactions, change, integration/disintegration is relevant. So, in a sense, all technology has its impact on some aspect of human existence that may play out into broad change or retreat.

Stefan: How would you evaluate technology's role for our understanding of continuity and change? How does it fit into this nexus of continuity and change and does it actually help us to understand this better?

Yale: It is absolutely critical to understand it better. But it really opens up a Pandora's box of fascinating questions. There are a number of important things that need to be said and/or asked here: One is to what extent should we think of technology itself as a "driver" of change or "merely" a "facilitator" or "enabler" of change. Let me illustrate this by mentioning a longstanding argument I had (that we both enjoyed) with my friend and former Rutgers colleague Richard Langhorne, who some years back wrote a well-known book on *The Coming of Globalization*, which (like Buzan's and Lawson's 2015 volume on *The Global Transformation*) put the spotlight on the "long nineteenth century." Richard always insisted that technology itself was the driver, that if there were technical advances, they would be used and that that in itself was the best explanation for change. I always countered that it was usually the *pre-existence* of a perceived need for technology that precipitated inventions in the first instance (well, some of them happened accidentially but not most) and then the full extent to which new inventions will actually be put to use is never entirely clear –and even less clear what the likely consequences of their use will be. I like to keep human agency front and center in the picture—and in a collective sense, general economic and social forces–not the science or engineering per se. These interact and create continuity or change or even retreat (change in the opposite direction). But I would not give technology too much credit in and of itself.

Let me make four other key points. First, technology can have multiple applications and, partly for this reason, unintended consequences. The Internet is a classic case in point. Originally developed by scientists, then funded by the US Department of Defense, obviously for reasons of security and warfare. But almost no one could have anticipated that the Internet would have become the global communications behemoth it is today—nor the threat it and social media have become to government control of information—nor the backlash that democratic governments have had to face because of their attempts to "listen in" on supposedly private networks—nor the foundation laid for the expansion of banks and multinational corporations around the world. Because they could much more easily communicate important information across distances and did not have to depend on telegraph or slow mail or even telephone. If you can't communicate easily, you can't run foreign operations very well. And this just opened up all kinds of possibilities. It has for example profoundly impacted travel agencies' monopoly on travel reservations—the use of the Internet to maintain social cohesion of ethnic minorities resident in foreign countries—the vulnerability of national information grids to cyber-warfare—and the issue of U.S. control of Internet regulation and broader governance issues.

A second point about technology illustrated by the Internet example is the fact that technological development tends to spin out of control well in advance of attempts to fashion and agree on any sort of governance or modest regulatory regimes. Space law is another early and continuing area of concern. Not to mention the tortuous history of regulation of nuclear weaponry and other forms of arms control. This is typical of the development of norms and law at both the national and international levels. Invention, acceptance, and practice tend to come well before recognition of the need for order and more formal rules.

Third, we should recognize that almost every positive technological development also has a downside. For instance, horrific environmental pollution caused by industrial development chocking Chinese cities and even the effect of economic "growth," in a generic sense, on the environment. Or consider the impact of advances in medical technology that allow people to live longer and, by doing so, place a burden on younger generations. Or, especially given the anti-trade rhetoric in the current U.S. presidential campaign, we should be revisiting the old argument about the relationship between automation and jobs. Many of the jobs lost by US workers, that are frequently attributed to off-shoring coupled with foreign competition, have, in fact, simply fallen victim to automation at home. Moreover, although some analysts now point to a significant trend toward re-shoring to the United States of some production—in part because of rising labor costs and dissatisfaction with other conditions abroad–the fact is that many of those work-slots (let us not call them jobs) are in fact going to machines. For their part, even the Chinese, to save rising labor costs in a highly competitive regional environment, are making greater use of automation technology.

Fourth, I think it is absolutely essential to analyze both technological change and attempts at norm-building and more mundane regulatory control with the longest possible historical perspective: the development of sea-going vessels starting with the earliest dugout canoes crossing short distances in the Mediterranean to sophisticated modern warships and submarines and giant container vessels. Rocketry from early primitive gunpowder versions in China through the German efforts in the closing days of WWII, to nuclear missiles during the era of Mutual Assured Destruction, to the improved nuclear delivery systems now under development by the U.S., Russia, and others that may restore a credible and thus destabilizing first strike capability, to the exploration of space, the placing of a skyfull of satellites, and on and on. Communication from bonfires and smoke signals to nearly instantaneous Wi-Fi connections across most of the planet, and so forth. If you look at technology that way, beginning really early on, you can see the trajectories and also the linkages across those trajectories that are truly fascinating.

Returning to a postinternational perspective, I have been struck recently by the possible impact of technology on the fundamental character of globalization. We may be seeing some major changes here. As we mentioned, the Internet almost in itself made possible the far-flung operations of multinational corporations that no longer had to use telegraph or wait for sea-mail or airmail to deliver their important messages to their managerial home base. Today it may be no accident that the recent deep recession in international trade has coincided with a massive upsurge in cross-border data transmissions. In fact I have some statistics on this. The flows of finance and goods and services fell from a peak of 53% of global output in 2007 to 37% of global output in 2014. That's a major drop. During some of this recent period, cross border data flows were picking up. Between 2013 and 2015 cross border data flows doubled. And they are adding another 30% by this year. And according to most estimates, by the end of 2016, companies and individuals will send 20 times more data than in 2008. This is developing in an almost geometrical progression. So, you begin to ask questions. For example, with the advent of ever-more-sophisticated forms of 3-D printing, it may become increasingly possible for countries to produce a greater variety of goods at home rather than relying on so many imports. Thus advances in big-data technology may be just the ticket for the new protectionist era that, for example, the anti-free trade stance of all the front-runner candidates in the current US presidential race might suggest lies right around the corner. Going back to what we (Ferguson/Mansbach) talked about some years ago in the "The Return of Borders to a Borderless World?", we argued in that book that we could see some trends in that direction, but thought it would be very difficult to reverse trade flows. However, if you project some of these trends much further into the future, it may be possible for countries that have a large domestic market (e.g.: China or the US) to be less open to the world and less engaged and still produce and get some of the same benefits, in a less tangible and more data-based way. *Stefan: What do you think could be prospective future research areas that might be interesting to explore within the technology-change-continuity nexus?*

Yale: We are very far indeed from fully understanding the role of technology affecting continuity and change in today's world. It is an area that needs a lot more research and thought, because the technology-related possibilities are often so mind-boggling. This also includes more empirical work. So, almost all the observations I have made are just that—observations—and beg further research: (a) the multiple applications and hence possible unintended consequences of specific new technological developments and those that are even now on the horizon (e.g.: drones); (b) the practical and normative upsides and downsides of past, present, and likely future technological developments on human life;

(c) the regulatory and legal regimes that need to be developed to mitigate those foreseen and unforeseen downsides—and the political feasibility of their being adopted; (d) the historical genealogy of specific inventions; and (e) the precise effect of big data on current and likely future global trade.

S.F.: Thank you so much for this interview Yale.

CHAPTER 12

Technology: From the Background to Opportunity

Leonie Tanczer in conversation with Barry Buzan

Abstract This conversation takes place between Leonie Tanczer of University College London and Barry Buzan, Professor Emeritus of International Relations at the London School of Economics. Buzan is a central figure of the Copenhagen School and Regional Security Complex Theory. He works currently with the English School approach to International Relations (IR), which is based on the understanding that there is not just an international system, but also an international society. The chapter discusses Buzan's experience with technology. It highlights how the material side of technology is quite conventional in IR and how we might best understand technologies as the instruments and the knowledge behind tools that enhance the capabilities of both humans and animals to do things. As Buzan points out, we can hardly

L. Tanczer (✉)
University College London, London, UK

B. Buzan
London School of Economics, London, UK

C. Kaltofen et al. (eds.), *Technologies of International Relations*,
https://doi.org/10.1007/978-3-319-97418-7_12

conceptualize anything in IR without thinking about the background technological conditions. Furthermore, we should go beyond seeing technology as a 'sector', rather conceptualizing it as a variable that affects all domains of the international system.

Keywords International society · English school · Change · Internet · International history

Barry Buzan is Professor Emeritus of International Relations at the London School of Economics and honorary professor at the University of Copenhagen, Jilin University and China Foreign Affairs University. He is probably best known for his publications *People, States & Fear: The National Security Problem in International Relations* (1983) and *Security: A New Framework for Analysis* (1998) which he published together with Ole Wæver and Jaap De Wilde. Buzan is a central figure of the Copenhagen School and Regional Security Complex Theory. He is currently very engaged with the English School approach which is based on the understanding that there is not just an international system, but also an international society.

> *Leonie: What understanding of the term and concept 'technology' do you have, or is it, similarly to the concept of security, basically an "essentially contested concept"?*
>
> *Barry:* I am sure it is an "essentially contested concept", because most things in the social sciences are. My view on the material side of technology is probably quite conventional. I would see technologies as the instruments and the knowledge behind tools that enhance the capabilities of both humans and animals to do things. Thus, I would favour a very broad understanding of technology. But I would also think of there being things like social technologies. This was something that Richard Little and I argued in *International Systems in World History* (2000). In this publication we refer to various kinds of social institutions like money, intergovernmental organisations or international law. In a sense these are also technologies which fit that general definition in the same way.
>
> *Leonie: You are a key voice within IR scholarship. On reflection, how do you feel technology has been present in your work and your core publications?*

Barry: Since I was knee-high, I was interested in war, military, and weapons. This interest got me into IR, and particularly 'Strategic Studies' as it was called back then. It made technology very central to my early work, but at this stage mostly in a military sense. You find technology in *People, States and Fear* (1998), when it discusses the defence and the power-security dilemmas; you find it in the *An Introduction to Strategic Studies* (1987), which is entirely about military technology; and you find it again in my book with Lena Hansen on *The Evolution of International Security Studies* (2009), especially when referring to nuclear weapons and nuclear proliferation.

While technology has not played a particularly central role in my work on the English School, it is far more present in my engagement with Regional Security Complex Theory. If you read my publications closely, there is a caveat which says: "This theory depends on territoriality continuing to have political and economic and military significance". If territoriality no longer has this significance—in other words—if one moved into a globalised world with very high connectivity and interaction, then Regional Security Complex Theory would not work. So in that sense, the theory is dependent on what I have elsewhere called "interaction capacity". It is the capacity to move people, things, goods, and ideas around the system. This is to some extent dependent on geography, but in modern times it is dependent on technology: are you in a world of sailing ships and horses or are you in a world of steamships, telegraphs, Internet and such like. So technology is a major background variable in that.

For me technology is also a huge mediating factor in how we think about international systems. This can be seen in *The Logic of Anarchy* (1993) that I wrote together with Richard Little and Charles Jones as well as my recent publication with George Lawson on *The Global Transformation* (2015). The latter is looking at the 19th century and examines a series of interlinked transformations, of which technology is one. The 19th century changed the world and laid down the foundations of the modern international system as we know it. It is a system that is created by new technologies of transportation and communication. So in that sense, technology was my starting interest in the subject and it has been there in a variety of ways ever since.

Leonie: Where do you see a relationship between technologies and classical IR concepts such as power, security, anarchy, and global order?

Barry: I would say you can hardly conceptualise anything in IRs without thinking about the background technological conditions. If you look at military concepts, you are immediately thinking of technology. You are then thinking of what kind of communication systems you have got, at what speed, and over what distances. You are also thinking of what kind of destructive capacities you have got, at what speed, and over what distances. Technologies frame the conditions of war and the dynamics of a system. So in that, sense all of these concepts rest on a set of technological preconditions.

Leonie: *In your film reading for 'Millennium: Journal of International Studies' you examine the international relations of Star Trek and Battlestar Galactica. You focus on how the United States views its own destiny and its relationship to technology and place in the universe. I like the idea of using science fiction to study dynamics of IR and technology. Do you see more potential for that?*

Barry: Yes and I am aware that other people are doing this already. For example, Stephen Benedict Dyson has written a short introductory text to IR based on the analysis of Star Trek, Battlestar Galactica, and Game of Thrones. *Otherworldly Politics* (2015) was the result of Dyson's teaching experience in Beijing. He used references to such shows to explain IR concepts to his students, making it easier for his class to relate to what he was talking about. Thus, as a teaching tool such investigations are certainly promising.

Besides, science fiction—or rather the practice of reasoning about future scenarios—is also a useful opportunity to start thinking about the unthinkable. For example, Dan Deudney and I are currently trying to reflect on the consequences of artificial intelligence (AI) for IR. At one stage in the future—some well-informed people are thinking in 2035 or maybe 2045—there will be a form of human created intelligence that is superior or equally superior to the Mark 1 human being i.e., you and me. The question is: what happens then? IR has not engaged with this kind of question yet. Indeed, it is a little bit like how IR has failed to engage with the scenario of the aftermath of a nuclear war. IR took you all the way up to the point of nuclear war and said: this might happen, this might happen, this might happen. But that was the end of the story.

I would say the only way you can think about the potential outcomes of such incidents is using some of the tools of science fiction. It is this systematic thinking about scenarios that would be applicable to quite a lot of the—what you might want to call—big threats that are looming in the not too distance future, may it be global disease or the rising sea levels. At this moment of time, IR does not think of that. It is very much focused on small and medium problems and on assumptions about continuities. It has not really started to reflect about big disjunctures. It seems to me that science fiction provides here an opportunity for IR. Besides, it might also bring a bit of fun to the field!

Leonie: *In your book Security: A New Framework for Analysis (1998) you are broadening security by outlining five 'sectors' of security. You are thereby*

referring to military, political, economic, social, and environmental secu-
rity. I am aware that there is a lot of discussion about the expansion of
these sectors. I was wondering if you would see the Internet, or cyberspace,
as being one of these new security sectors. Would it qualify as a functional
differentiation? Is there consequently a way of speaking of, for example, a
"technological sector"? Or is encompassed in the other five or more sectors?
Barry: I have spent a lot of time trying to theorise sectors, especially in
work with Mathias Albert. Albert, due to his sociological background,
noticed that the notion of sector was similar to the notion of functional
differentiation prevalent within sociology. The original five sectors—
as it does say in the book—are empirically derived. In other words, if you
looked at the literature on international security at that time, those five
sectors would come up all the time. These sectors were not derived from
any theory, and I have failed to find a way to do that. Instead, it was
a purely empirical derivation, though they did embody a certain functional differentiation that characterised them. One could use the same
empirical method nowadays to try to assess what is being talked about
in the literature. There is therefore absolutely nothing set in stone about
sectors.

There is—as you rightly pointed out—a much talk about cyberspace being one of these new sectors. And I acknowledge that if we
were writing that book nowadays, we would probably have picked that
up empirically. Yet, it is not quite a sector in the functional sense than
the other five sectors are. You could make the argument—as you hint
in your comment—that cyberspace is in some sense an aspect of the
other sectors. There is a military, political, economic, and social aspect
to it. It does, however, have some distinctive characteristics of its own.
Thus, if you just take sectors as meaning: the things to which security
is attached—for example cybersecurity, food security etcetera—maybe
cyberspace should also be thought of as a new sector.

Nevertheless, I do not think that there is a technology sector. I think
that would be going too far. Technology is a variable that affects—for
the reasons I have been talking about earlier—all the other sectors in
different ways. The attempt to make this a sector in its own right does
therefore not resonate with me. That being said, I would not stand in
your way if you want to try to do it. You would have to make the case
for it and I think that would be a bit difficult. But prove me wrong!

Leonie: You and Ole Wæver have used the five sectors framework to analyse
how and why different issues and sectors are securitized in different regions
of the world. Would you see any aspects of technological issues such as the
Internet being securitised?

Barry: It seems to me the only sensible way to answer this question is to say that anything can be securitised. Securitisation theory is based on constructivism. It does not require that there is an actual, real threat. It simply depends on what can be successfully constructed as a threat. In that sense, there is nothing distinctive about technology. Its securitisation will depend on the local conditions. Once you start looking at the local conditions, some things become easier to securitise than others, for example, due to cultural or historical reasons, or because of the nature of international society.

Leonie: Across your work you emphasise the importance of re-connecting IR with history and sociology and refer to the importance of the 19th century for the character of international society to this day. However, looking at the way society is currently developing, would you not say there will be a need for IR to connect with engineers and computer scientists soon as well?

Barry: I would not underestimate the difficulty of doing that. You have probably been in academia long enough to know how quickly academics seal themselves off into little discursive entities.

They share a certain kind of jargon and vocabulary, they publish in certain journals, and they talk to each other and not to anybody else. This happens within departments, let alone between them. So there is what one might describe as a language problem in doing that. It is like the problem of people trained in political science talking to economists. Unless you learn the language of economics, you cannot talk to economists. So it is actually very difficult to do that.

Indeed, my preferred route would be to make sure that all academic disciplines encourage generalist as well as specialists. You need people who know something about everything, but not much about anything in huge depth. That is the way I have organised my own career. The thing that attracted me about IR was that you could be interested in everything and you did not have to acquire a discipline. You could dabble in all kinds of things. However, the name of the game is then: can you provide added value to those specialists who know a lot about some small patch of time and space? And the answer is yes. Increasingly, as we get more and more specialised, we need generalists. Nonetheless, we are not doing a very good job in producing them.

Leonie: But would you say – coming back to the topic of this conversation – because of the importance of technologies and specifically digital technologies, IR has to or will have to move there?

Barry: Yes, because most books on the matter are written by engineers. These are people who know the tech-side of things, but they know nothing about the context of technologies. So these books, while interesting, are very narrow. Technologists need generalists too. These generalists are then able to hook the discussions into wider debates in the social sciences. For the problems that society is facing, we require lots of specialist knowledge being linked together. The less this knowledge is linked together, the less we are able to answer some of the big, urgent questions in front of us. How we are going to do that is not clear, because there is this structural problem there. The internal dynamic of academic life tends to push academics more and more towards specialisation. This specialist imperative of an academic career is standing in the way of the need to tie expertise together.

Leonie: Perhaps IR is a good discipline for doing that, because it has always been a cross-disciplinary field. On the basis this particular status, what do you see as the most significant challenges for IRs in the context of technological change?

Barry: Keeping up is a pretty big challenge. There is an aspect of what I once called "hectic empiricism". If you were involved in Strategic Studies during the Cold War, maybe eighty or ninety per cent of the energy in that field was taken up with trying to keep abreast of what the new technological developments were and thinking about how they would play into deterrence logic, escalation logic, and arms control. Keeping up with these developments was pretty much absorbing all of the energy that was available. There was relatively little left over for trying to think outside the box and think about bigger questions. And it seems to me that we are still in that game. The cyber theme is evolving at a stupendous rate and so are other things such as AI or biotech. Trying to keep up with all of that is challenging, especially if we are aiming to get ahead of it in some way. As I said earlier: we should be thinking about the unthinkable, thinking about the scenarios of the day after some big change. If we were able to do that, we would be doing pretty well.

Leonie: Conversely, what do you see as the most significant opportunities for IRs in the information age we are currently living in?

Barry: The information age provides opportunities for the globalisation of IR. IR as a discipline is, has been, and will probably continue to be far too much based on Western history. We all pretend that Western history is world history and that Western political theory is political theory – but it is not. It seems to me that there is a real need to bring other people's histories and other people's political theories into thinking about IR in a more global way. The technologically connected world and technologies underlying interaction capacities provides opportunities to achieve this. Technologies will consequently facilitate this overdue movement which is an action that needs to happen as quickly as it can.

Leonie: Lastly – a more personal question – what are the five key technologies you could no longer live without and that help you in your day to day life, but especially in your scholarly work.

Barry: 'Not living my life without' is a tough criterion for me. I could get by with the pre-digital tools I used in the first half of my life, and live without computing and word processing, although it would be a somewhat impoverished existence. But that said: email is absolutely crucial to me: most of my collaborative work is now done that way. Second, word processing. I got my first word processor in the early 1980s and it transformed my life and also changed the way I wrote. Web browsing is another technology I, as anybody else, am dependent on. For example, if I give money to charities it is generally to Wikipedia. In my growing old age and as my eye sight is deteriorating, on-screen reading is also becoming vital for me. This technology allows me to control the light and the font size. My number five key technology would be my Pavoni Coffee machine. It produces brilliant coffee which keeps me going and is an important fuel for my activities!

Leonie: Professor Buzan, thank you for taking the time talking to me.

Works Referenced

Buzan, B. (1998). *People, states and fear.* Boulder, CO: Lynne Rienner Publishers.

Buzan, B., & Lawson, G. (2015). *The global transformation: History, modernity and the making of international relations.* Cambridge: Cambridge University Press.

Buzan, B., Jones, C. A., & Little, R. (1993). *The logic of anarchy: Neorealism to structural realism.* New York: Columbia University Press.

Buzan, B., Wæver, O., & De Wilde, J. (1998). *Security: A new framework for analysis.* Boulder, CO: Lynne Rienner Publishers.

CHAPTER 13

'New Technologies': Questions of Agency, Responsibility, and Luck

Sarah Logan in conversation with Toni Erskine

Abstract In this wide-ranging conversation Professor Erskine details the emergence of her interest in new technologies and their impact on international ethics and politics. She explains how a research interest in the early stages of her career in the ethics and norms of war led via work on institutional moral agency to her current research on artificial intelligence. She outlines the key questions animating her work in the context of AI, robots and moral agency and moral standing. She goes on to explain how technology has become increasingly foregrounded in her work and discusses the impact of technology on International Relations.

Keywords Artificial intelligence · Robots · Ethics · War · Norms · Intelligence · Technology · Agency · International relations

S. Logan (✉)
University of New South Wales, Sydney, NSW, Australia

T. Erskine
Coral Bell School of Asia Pacific Affairs, Australian National University, Canberra, ACT, Australia

C. Kaltofen et al. (eds.), *Technologies of International Relations*, https://doi.org/10.1007/978-3-319-97418-7_13

Sarah: Thank you very much for your time, Professor Erskine. To start with, I wanted to clarify the context and the concepts involved in this discussion. So, it may seem like a very broad question, but I wondered if I could begin by asking what the term "technology" means to you?

Toni: "Technology" is a common, but amorphous, term in our everyday, twenty-first-century vernacular. I labelled a broad, collaborative research project that I recently initiated 'New Technologies and the Ethics of War', and then realized (when questioned by an applicant for a postdoctoral position within the project) that I hadn't taken the time to define 'technologies'. I took for granted that everyone would understand what I meant. (This was out of character; I tend to be somewhat obsessive about defining concepts!) I realized that 'technology' is more difficult to pin down than I had assumed. There are (at least) two ways that the term is used. There is also, I think, a sentiment that currently accompanies this term, which goes beyond any straightforward definition.

The *Oxford English Dictionary* defines technology as "the application of scientific and technical knowledge for practical purposes" *or* "the machinery, equipment etc. developed from the practical application of such knowledge". I use the term in both senses—technical methods/processes and the artefacts thereby created—and was able to clarify that I had intended the latter in choosing the phrase 'new technologies' as part of my project title. My interest is in the tools, machines, devices, and equipment that are the product of our scientific and technical knowledge. In both the collaborative project that I was labelling and my own recent work, my focus has been on tools in the form of weapon systems. As my recent work has been specifically concerned with technologies that make use of artificial intelligence (AI)—so computers, machines, and robots with capacities to demonstrate or mimic intelligent behaviour—the modifier "new" is important. After all, despite how the term is often invoked (as necessarily new even without the modifier), and despite its relatively recent prominence, "technology" (as defined by the OED) has defined *every* era of human development.

In addition to where my use of the term fits alongside strict dictionary definitions, it is worth noting that there is a sense in which the word "technology", when uttered at the start of the twenty-first century, can also connote something else. It frequently seems to evoke a sense of mystery, or awe—sometimes even fear. This is because, I think, it is associated with processes and devices that most of us don't understand.

It also carries a sense of change, even revolution, in the way things are done. "Technology" is unsettling.

Sarah: And when you first began your career was your work influenced by thinking about technologies at all, or is this something that's developed over time?

Toni: No—not consciously anyway. I've only explicitly turned my attention to the political and ethical dimensions of "new technologies"—including so-called lethal autonomous weapon systems (LAWS), AI technologies more broadly, and cyber capacities—over the past few years. Nevertheless, from early in my career, beginning with my doctoral research, I've focused on the ethics of war and, specifically, how one understands and treats "the enemy". An important aspect of this has been a focus on norms surrounding who it is morally permissible to kill and how. This involved, in turn, some consideration of the nature of military technologies at different points in time (and here I mean anything from bows and arrows, catapults, and gun powder to chemical and biological weapons, submarines, nuclear arsenals, and unmanned aerial vehicles, or "drones"...). So, technologies *did* play a role in my research. But, to be honest, I tended not to address this concept directly and wouldn't have seen myself as 'thinking about technologies' at that point.

Sarah: What triggered your interest in "new technologies"?

Toni: My recent, explicit interest in 'new technologies' and their impact on international ethics and politics has been sparked by three things. First, I'm coming to the end of a long-term project on what I call 'institutional moral agency', which defends the idea that certain collectivities (such as states, intergovernmental organisations, and transnational corporations) can be moral agents in their own right, or, in other words, in a way that is not reducible to their individual human constituents. (By moral agents, I mean purposive agents to which we can reasonably assign moral responsibilities and apportion blame in the context of particular acts and omissions.) This position challenges the prevalent view that only individual human beings—like you and me—can be moral agents. It struck me that including such corporate entities within the class of moral agent might be understood to have implications for forms of AI. And this really worried me. Given the understanding of moral agency that I invoke to make the case that formal organisations qualify—namely, that moral agency requires sophisticated, integrated capacities for deliberation and action, but not other characteristic that we associate with being human—there is no obvious reason why some machines, computers, and robots could not also qualify at some point in the future. Yet, the idea that sophisticated *robots* could be considered the bearers of moral responsibilities seemed profoundly problematic. I wanted to figure out why this prospect made me so uneasy.

Second, I was appointed Acting Director of the nascent Australian Centre for Cyber Security at the University of New South Wales (UNSW) shortly after I arrived in Australia in 2013. The role involved drafting and submitting a detailed (and ultimately successful) proposal for the new centre and bringing together an interdisciplinary group of scholars. This was serendipity rather than research strategy, but it exposed me to a new area of inquiry, a new and wonderfully diverse community of scholars, a new set of pressing practical problems, and a new context in which to consider, for example, my long-standing interest in moral norms in international politics. Notably, one of the arguments that came out of my consideration of norms in cyberspace—in a piece co-authored with Madeline Carr (Erskine & Carr, 2016)—is that existing norms, which have emerged and evolved in the context of particular practices, such as conventional warfare, cannot simply be parachuted into the cyber realm. For me, there is also a more general principle that has come out of this work: that "new technologies", by their very nature, cannot be analysed by means of borrowed conceptual frameworks, second-hand assumptions, and weak analogies with what is already familiar. Perhaps this is why I see 'technology' as unsettling. The "new technologies" that I've been focusing on disrupt not only what we do, but how we think about the world—practice *and theory.*

Third, while a guest of the Centre for the Future of Intelligence at the University of Cambridge recently—to speak about my work on moral agency and what this might mean for thinking about responsibilities in relation to robots—I was invited to attend a separate workshop on a topic that I knew little about, also organized by the Centre. The title of the workshop was 'Psychometrics, Machine Learning and Democracy'. It sounded interesting. In fact, it was both riveting and deeply disturbing. The workshop addressed how social media sites, such as Facebook, can provide data points from which it is possible to construct sophisticated psychological profiles of individual citizens. These profiles can be used—indeed *have* been used—to produce targeted messaging in election campaigns, such as the 2016 US presidential election. I was struck by the possibility that emerging AI could provide new opportunities to manipulate citizens within democracies by targeting narratives designed to mould behaviour at particular individuals based on their personalities, fears, and vulnerabilities. Participants at the workshop addressed the case of the firm Cambridge Analytica, for example, and its efforts to influence elections in different parts of the world. We also explored the ethical problems involved in collecting data without informed consent, and of both using misinformation and targeting vulnerabilities to influence how citizens vote—or even to *suppress* voter turnout. We discussed the danger of such political manipulation when it is wielded by foreign powers. We lamented companies such as Facebook not having to disclose where the funding for this type of targeted messaging originates. And we addressed the problem of those targeted individuals being *unaware* that the stories they see are substantively different from, or even contradictory to, those presented by the same source to other citizens (with different psychological profiles)—in other words, that these stories were directed at them for a reason.

For me, the workshop was an epiphany. After a day of presentations and discussion amongst a small group of invited scholars, I left convinced of the importance of the subject matter—and worried that this topic wasn't being discussed broadly and with the urgency that it warrants. Interestingly, since the workshop, more news stories about cyber-threats to democracy *have* begun to appear, particularly regarding alleged Russian interference in the 2016 US election. But, it seems that we—as citizens and as academics—aren't taking the problem seriously enough. I realized after the Cambridge workshop that this was exactly what one of the postdoctoral fellows working with me, Dr Tim Aistrope, had already been trying to tell me about over the previous couple of months. Yet, I hadn't then understood what was involved, and hadn't registered the significance. (I should note that Tim was well ahead of the curve on this. He's the first IR scholar that I'm aware of who's been working on the problem of machine learning and what he calls 'cyber-enabled information operations', and I'm looking forward to learning more from his work.) I wonder if I had managed to sidestep the issue—until I was confronted with it directly in the intensive seminar—because I didn't understand the technology behind it, didn't understand what was possible, and somehow thought (wrongly!) that it wasn't relevant to my general research interests in international politics. Luckily, the pieces began to fit together for me after the workshop: what I'd learned recently about AI, my multidisciplinary exposure to cyber security, what Tim had told me about information warfare, my political science interest in voting, and my political theory interest in the nature of liberal democracies (including their vulnerabilities)... This is a topic that I'm determined to study further. And, this is something that I think individual citizens have a *responsibility* to be aware of and to educate themselves about. Such a responsibility must surely be even greater for academics studying politics and international relations.

Sarah: I guess that's what I want to talk a little bit more about—how your thinking about technology influences the direction of your work. And so, for example, if we look at your work on the ethics of intelligence gathering, were you thinking about the role of technology when you were writing that?[1] I mean, at the time you wrote it, which I think was 2003/2004, it wasn't really something that was on our radar—to use a technological term(!)

Toni: No, I don't think that my work at *that* point was—again, consciously—considering the role of technology. I touched on surveillance and forms of signals intelligence in examples of how one might morally assess different means of intelligence collection from three distinct ethical perspectives. Yet, I treated technology as a barely-visible backdrop against which I was asking questions about how we should and shouldn't treat fellow citizens, foreigners, and 'enemies'—and why.

Rather than considerations of technology influencing the direction of my work at that point of time, I think that it is more accurate to describe my current work on new technologies being influenced by insights gained from my previous research—when I wasn't really paying attention to technology directly. For example, I'm sure that my work in the 2004 *Intelligence and National Security* article that you mentioned has informed the ethical questions that I've begun to ask about psychologically targeted messaging in elections—regarding issues of privacy, transparency, indirect harm, and consent, for example. As I alluded to, I've been confronted by the reality and significance of new technologies through a variety of circumstances and have gradually realized that my work across a range of areas—moral agency, responsibility, norms, foreign interference/intervention—wouldn't be complete if I didn't also explore the ethical and political impact of particular technologies directly.

So, my current focus on specific new technologies has allowed me to pursue unexplored avenues related to my previous research that I'm only now aware of. For example, there are at least three sets of questions that I'm asking in relation to my most recent work on artificial intelligence, which I realize—thanks in large part to this conversation!—have been influenced by my previous work. One set of questions relates to moral agency. Can forms of AI—computers, robots, machines—be moral agents? And, does exploring this possibility help us to understand the individual human and corporate manifestations of moral agency? Another set of questions relates to the separate puzzle of moral standing. Are AI entities morally considerable? Do we have duties towards them? Could it ever make sense to say that forms of AI *themselves* have non-derivative moral rights? And a final set of questions involves our perceptions of our own moral responsibilities. How might the existence of sophisticated forms of AI affect how we understand our individual and corporate responsibilities in international politics? Do we see our responsibilities as somehow diminished if some AI entities *appear* to have capacities to deliberate and act on their own? The first and final sets of questions are prompted, I think, by themes in my current book project, *Locating Responsibility: Institutional Moral Agency and International Relations.* The second line of inquiry has its genesis in an earlier monograph (Erskine, 2008). There is at least some continuity in the types of questions that I'm asking!

Note that I'm not saying that I can draw on previous theories or conceptual frameworks to *answer* these new questions. As I mentioned a few moments ago, I don't think that this would be possible. Rather, my earlier work has influenced the problems that I'm posing—what I'm particularly curious about—and has thrown into sharp relief the conceptual disjunctures and challenges that accompany trying to analyse specific new technologies.

Sarah: Thank you. You just mentioned some of your more recent work on artificial intelligence, including the idea of robots as moral agents. I wonder if you could talk a little bit about what has come out of your work in this area so far—and what, if anything, has really surprised you.

Toni: There are, for me, a few notable things that have come out of my work in this area so far. The first is that after my initial (unexamined) denial, I'm willing to accept that what I have tentatively called 'simulated moral agents'—forms of AI that could coherently be assigned moral responsibilities and apportioned blame—are at least *conceivable.* I don't think that any exist yet. And, I argue that, were they to emerge, they would be very different from both their individual flesh-and-blood and corporate counterparts. But they are, I think, possible—and what their existence would mean and how exactly it would manifest are important problems to consider.

Another realization—one that I'm surprised and intrigued by—is the extent to which we tend to athropomorphize forms of artificial intelligence, and—I think, problematically—assume that they are morally considerable, that *we* may have moral responsibilities towards *them*. I came across a story while working on a recent article on AI and moral responsibility in war that I think illustrates well the tendency of individual human actors to empathize with some robots. A military robot the length of a small adult, with a body resembling a "stick-insect", was designed to tread on and destroy landmines, thereby protecting soldiers from risk. According to an account published in the *Washington Post*,[2] an incredibly successful test run (from the perspective of the engineer who designed it) saw the robot repeatedly encounter a mine, detonate it, lose a 'limb', and then pick itself up and proceed forward on its remaining legs. It did this until it only had one leg left—and still somehow managed to pull itself forward. However, at this point, according to the published account, the colonel overseeing the exercise could take no more and called an end to the test. He exclaimed that it could not continue because it was '*inhumane*'. 'Inhumane'? Think about that for a moment. The term suggests a lack of compassion for suffering. The colonel's reported outburst is amusing—but also startling. Perhaps it's better to imagine that some robots have moral rights than to anthropomorphize them and treat them badly. After all, there might be a risk that this would encourage us to treat actual sentient beings in a similar manner. Yet, if it's so easy to misperceive military robots as things that can suffer, might it not be similarly easy to slip into the mindset that the most sophisticated among them are also moral agents?

Finally, following on from this last point, regardless of whether AI entities can ever be moral agents, I've been struck in my recent work by what I think is a neglected danger: that the tendency to *perceive* them as such already informs how the individual human agents that act alongside them view their own moral responsibilities. If there's some sense that a military robot, for example, is actually making a decision and acting without human oversight, does that let the individual human agent acting alongside it off the moral hook? I think that the problematic perception that forms of AI can act genuinely autonomously allows us—as individual human agents—to see our own moral responsibilities as somehow (and rather conveniently) diminished. This could have catastrophic consequences.

Sarah: One of the things that's striking about forms of artificial intelligence as moral agents is the idea of deliberation, and how we understand this. Are you able to talk about understanding deliberation in the context of artificial intelligence?

Toni: In very simple terms, to qualify as a moral agent, an entity must have both the capacity to deliberate over possible courses of action and their consequences and the capacity to act on the basis of this deliberation. When it comes to forms of artificial intelligence, I argue that the main obstacle to qualifying as a moral agent is autonomy, or the capacity for the entity in question to deliberate and act on its own.

Unpacking what it would mean for a robot to deliberate and act autonomously is an important exercise. Military robots are often described as being 'autonomous', but this is in the very weak sense of being programmed, for example, to identify and fire on specific types of targets independently. So, the Phalanx, a navy missile system, uses a computerized radar system to search for, identify, and engage targets without human operators. This is, indeed, a very sophisticated piece of technology—yet, it does not display 'autonomy' in the sense required for a machine to qualify as a moral agent. This must be autonomy of a very different kind.

As I've argued in a recent paper (Erskine, 2017), for a machine to be considered a moral agent, it must be able to do a number of different things without human control. It must be able to access and interpret information. It must be able to draw on this analysis in order to decide on a course of action. It must be able to act on the basis of this decision. And, it must be able to *learn*, by which I mean adjust its decision-making processes and reconfigure its preferences based on experience. This final criterion that I'm proposing is important. The plans and preferences of AI entities remain those of individual human engineers, programmers, and operators unless the machine itself can reflect on, and alter, both its decision-making processes and plans and preferences independently. And there's the rub. I want to suggest that military robots, for example, could only reasonably be understood as bearers of moral responsibilities in their own right if they were able to contemplate the moral guidelines with which they were programmed, revise them based on this deliberation—and, crucially, thereby choose to do other than what their programmers intended. If one accepts this final criterion that I'm proposing, there are currently no machines that qualify as moral agents. I'm interested in whether this criterion could ever be met when it comes to AI technologies.

Sarah: And that brings me to my next point, which is about the complexity of advanced technologies. Particularly in the field that you're working in, one of the striking elements of that is that things are developing extremely quickly. And it feels as though—and, you know, sometimes we like to overemphasise this, I think—as though we don't know where it's going to end up. Do you think it's possible or useful to think about technological complexity and the speed of technological development? Or is it an issue at all?

Toni: I think that the speed and complexity of technological development—and the profound uncertainty of where we might end up as a result—are fundamentally important. Keeping with the example of military robots, we might note that Elon Musk recently lamented in a series of tweets that the main risk of another world war lies not in the actions of rogue states or violent non-state actors, but, rather, in the possibility that intelligent machines might become truly autonomous and be able to make decisions for themselves. *They* might decide, he suggested, that the best way to achieve victory is with a pre-emptive strike.

This may seem an unlikely scenario, but at its heart is the idea—the fear—that complex technologies such as sophisticated military robots are developing so quickly, and with such uncertain potential, that we risk losing control of what we are creating—and what these new technologies will ultimately do. This is the concern behind the move to ban the development of so-called 'killer robots', for example.

Sarah: And do you think this idea of managing uncertainty is unique to this problem of technological complexity? Or do you think it emerges elsewhere as well?

Toni: No—I don't think that 'managing uncertainty' is unique to the problem of technological complexity. Given the intricate structure of our highly interdependent global system, causal chains are far from straightforward—and impossible to predict. Making decisions under conditions of uncertainty is therefore a formidable challenge in international relations generally at the start of the twenty-first century. This is a challenge that I have thought about recently in the context of retrospective moral judgements—so with respect to praising or blaming the decisions that agents have made to pursue or refrain from particular courses of action when the consequence of doing so is unclear. I've been particularly interested in the philosophical idea of 'moral luck', introduced Bernard Williams and extended by Thomas Nagel in the late 1970s.

To give a too-brief account of a difficult idea, "moral luck" is the purposely paradoxical idea that our moral judgement is unavoidably informed, at least in part, by chance and contingency. It is paradoxical because morality is supposed to be immune from luck. When we judge agents for their acts and omissions, we are meant to take account only of what is under their control—so what they intended to do, or the foreseeable but unintended consequences that resulted from negligence or recklessness. When one makes a decision under conditions of uncertainty, though, there are often outcomes of that decision that are not what one intended, and that also did not result from negligence or recklessness, but, rather, were influenced to some degree by luck. Williams and Nagel suggested that a variation on moral luck—what Nagel calls 'resultant luck'—occurs if a decision is praised or blamed on the basis of its outcome, even though the outcome was not entirely within the agent's control.

I have explored—and worried about—this idea in relation to retrospective evaluations of decisions to intervene militarily for human protection purposes. I think that it is tempting to judge an agent's decision to intervene based on the outcome of the military action. Yet, this is, I think, a good example of moral luck and, as such, is necessarily problematic. The assessment of the decision itself shouldn't depend on luck—either good or bad—in this way. Rather, backward-looking evaluations of a decision to act or refrain from acting should be judged according to what it was possible for the agent to know at the time. In other words, this variation on moral luck is an *error* of judgement—and is avoidable. Nevertheless, I've also argued that one can take a lesson from this notion of resultant moral luck. It usefully gestures towards how we should behave when we are making potentially consequential decisions under conditions of uncertainty (Erskine, 2018). Namely, it reinforces the imperative to act only when it is possible to offer a fully informed, compelling justification of the decision to do so, which heeds clear precautionary principles and will withstand any outcome.

So, in sum (after a rather round-about response!), complexity and uncertainty are, I think, common challenges in international relations. And, ways that I've tried to think about decision-making under conditions of uncertainty in international relations more generally might be useful when considering uncertainty arising from technological complexity. This is something that I need to consider more, but it strikes me that a lesson gleaned from the idea of resultant moral luck is relevant to thinking about decisions that we make regarding the development and use of new technologies. Namely, when the impact of our decisions to develop and use new technologies in particular ways is unclear, then the onus should be on the agents making decisions to exercise caution in light of what is known, and with recognition of what can't be known due to technological complexity. This doesn't mean blaming developers for decisions based on outcomes that couldn't have possibly been foreseen, but, rather, assessing agents' decisions about the development and use of technology based on how these decisions were made and the expectation that agents should take into account acute uncertainty.

Sarah: And what about the companies that design this technology? How should we think about the responsibilities of the corporations that produce technologies that may cause harm, even if they're removed from the eventual acts and outcomes that result from their products?

Toni: I think that a really interesting—and difficult—challenge involves assessing the responsibilities of corporate (and individual) developers of technologies that are used for purposes not originally intended. It's not always obvious when developing new technologies how they will be used in the future, or even what their potential uses might be. Blaming corporate (and individual) developers of technologies for their future harmful consequences may be unfair if these effects couldn't have been foreseen. (And here we are back to the possibility of moral luck in relation to blaming agents retrospectively for even well-intentioned and carefully considered decisions made under uncertainty.) There has been talk of the 'weaponization of Facebook', for example. This sounds extreme, and perhaps even alarmist, but if we think of perhaps highly consequential 'information warfare' by foreign powers (and here again I'm thinking of discussions with Tim Aistrope), then this label doesn't seem so far-fetched. Yet, whatever the ambitions were for Facebook when it was first being developed—and I'm sure there were many—I don't think that its 'weaponization' was one of them. Not only was this (presumably) unintended by its designers, but I'm not sure that it could have been foreseen. It therefore seems problematic to blame the original developers of this social media platform—or indeed the pioneers of psychometrics, now being used in combination with big data and AI—for its malign use by foreign powers to influence elections through targeted 'propaganda ads'.

Nevertheless—as I've argued in my work on institutional moral agency—corporations *do* have moral responsibilities, and, moreover, can be blamed for negligence and recklessness, in addition to intentional harm. (At this point, I should note that an early motto at Facebook—'Move fast and break things'—does give me reason to pause in relation to the previous example and Facebook's potential culpability for the platform's subsequent malign use. Rapid creation seems to have been given explicit priority, at least for a time, over consideration of consequences.) If it's clear that some technology being developed is likely to do more harm than good, then the corporation does have a responsibility to self-regulate what it's doing regardless of questions of legality—and perhaps refrain from developing certain types of technologies altogether. (Note here that what are colloquially called 'killer robots', or lethal autonomous weapons systems, have purposely harmful functions, and attendant risks, and could never be seen to be as innocuous to their creators as Facebook may have seemed to theirs.)

Ultimately, I'm not sure that it's really practical to think that we can simply halt the development of certain types of technologies. What seems most important is to be aware of what is being developed, to start thinking about how it might be used, and then to talk about regulating its use once potentially harmful and malicious uses are apparent. And here Facebook as a corporate entity, along with the individuals within it in positions of power, cannot be excused for failing to mitigate and protect against the possible harm that we *now* know the social networking platform can be used to inflict.

Sarah: Thank you. I wondered if you could talk about the impact of technology on key IR concepts like order or power. Is that something that we can talk about?

Toni: For me, the key IR concept that technology current has the most profound impact on is *agency* and, by extension—for all the reasons I've highlighted,—*moral agency*.

Within IR theory there has been a long-held assumption that formal organisations like states are purposive agents. I've argued that we should understand them as moral agents as well. Addressing forms of artificial intelligence and saying, "perhaps these are also purposive agents that IR needs to be aware of" raises a whole host of important questions. And these questions are not limited to whether there's a new type of agent and a new type of moral agent or the horizon, but also whether examining this possibility helps us to better understand or even reconsider both individual human agents and institutional agents such as states. Separately, we also need to consider how our perception of new technologies affect how we understand ourselves as moral agents in particular contexts (when responsibility for an act or omission might *appear* to be borne by a machine)—and, moreover, whether new technologies contribute to already-complex causal structures that variously constrain and enable the moral agents within them. Technology will challenge, confuse, and force us to recalibrate how we understand agency and moral agency.

Sarah: Do you think there's enough focus on technology in IR? Is there room for more?

Toni: My short and adamant answer is 'no', we're definitely not paying enough attention to it—with, of course, some shining exceptions, particularly amongst a diverse new generation of IR scholars. It shouldn't just be some people who are working in this area—those specialising in new technologies and their impact on international politics. Such specialization is valuable, but if we think that it is enough to have some scholars working in a separate 'technology and IR' sub-field, then we have a problem. Everybody who is working on international politics needs to be aware of technological developments and needs to consider the impact of these developments—whether their work is concerned with democracies and voting, responsibility and war, communication, international security variously conceived, or simply how we understand agency. And, there must be collaboration between people working within IR with those working across other disciplines.

Sarah: So, as a final question, why do you think there is that gap? Is it just a matter of time lag?

Toni: I think that there are a few reasons for a gap between how many people in IR are actually taking the impact of technology seriously and the significance of the subject matter. There is a time lag as new technologies emerge and evolve and we—academics included—struggle to keep up. There is also a knowledge gap as we don't have the technical expertise to understand what is happening. (Hence my point just a moment ago that we need transdisciplinary collaboration.) And, with this lack of knowledge comes fear. Sidestepping issues associated with technology is more comfortable than trying to tackle what we can't completely grasp. Finally, there is perhaps no small amount of naiveté in our failure to appreciate the significance of the changes that are *already* directly affecting us.

Notes

1. T. Erskine, '"As Rays of Light to the Human Soul"? Moral Agents and Intelligence Gathering', *Intelligence and National Security*, Vol. 19, No. 2 (2004), 359–381.
2. J. Garreau, 'Bots on the Ground: In the Field of Battle (Or Even Above It), Robots are a Soldier's Best Friend', *The Washington Post* (2007), http://www.washingtonpost.com/wpdyn/content/article/2007/05/05/AR2007050501009.html.

Works Referenced

Erskine, T. (2008). *Embedded cosmopolitanism: Duties to strangers and enemies in a world of 'dislocated communities'*. Oxford: Oxford University Press.

Erskine, T. (2017). *Flesh-and-blood, corporate, robotic? Moral agents of restraint and the problem of misplaced responsibility in war*. Presented at the Artificial Agency and Collective Intelligence Workshop, Centre for the Future of Intelligence, University of Cambridge, 18 September 2017.

Erskine, T. (2018). Moral responsibility—and *luck?*—In international politics. In C. Brown & R. Eckersley (Eds.), *The Oxford handbook of international political theory*. Oxford: Oxford University Press.

Erskine, T., & Carr, M. (2016). Beyond "quasi-norms": The challenge and potential of engaging with norms in cyberspace. In *International cyber norms: Legal, policy & industry perspectives* (pp. 87–109). Tallinn: NATO CCD COE Publications.